Generation of Wealth

Generation of Wealth

The rise of Control Data and how it inspired an era of innovation and investment in the Upper Midwest

Donald M. Hall

Nodin Press

ISBN: 978-1-935666-63-9

Fourth printing 2015

Library of Congress Control Number: 2014931256

Design: John Toren

Published by
Nodin Press
5114 Cedar Lake Road
Minneapolis, MN 55416

To librarians everywhere, those soft-spoken servants of a public struggling to learn—and especially the librarians at the Minneapolis Central Branch of the Hennepin County Library, the James J. Hill Reference Library in St. Paul, and the Babbage Institute on the campus of the University of Minnesota, who helped me discover obscure caches of information for this book.

TIMELINE

Year	National /Local Events	Local Computer Events	Medical Device Events
1945	World War II ends		
1946		Engineering Research Associates (ERA) formed	
1949			Medtronic formed
1951		ERA sold to Remington Rand	
1957	Russian satellite Sputnik circles the earth	Control Data formed, stock offered at $1 a share	Portable pacemaker developed
1959			Medtronic public financing
1961	First man in space, 96 local stock offerings	Control Data stock at $101.50	
1962	Local market deflates		Medtronic offered for sale
1964		IBM announces paper machines	
1966		Key employees resign from Control Data	
1968	Widespread social unrest nationwide, 893 local stock applications, stock trading hours reduced	Control Data stock at peak, it acquires Comm. Credit, files suit against IBM, and initiates "social concerns"	
1969		Computer price "unbundling"	
1972		Cray Research formed	Cardiac Pacemakers formed
1973	Watergate, Arab oil embargo	Control Data – IBM law suit settled	
1974	Pres. Nixon resigns, Bear market low		
1976		Cray Research public offering	St. Jude Medical formed
1978			Cardiac Pacemakers acquired
1981		Seymour Cray becomes an independent contractor	
1987		Cray Research stock peaks	
1989		Cray Computer formed	
1992		Control Data becomes Ceridian and Control Data Systems	
1995		Cray Computer files bankruptcy	
1996		Cray Research is acquired	
1997		Control Data Systems is acquired	
2007		Ceridian is acquired	
2009	Bear market low		Medtronic and St. Jude are industry leaders

Contents

Foreword

This book captures the essence of an era, a time of great creativity when capital was available to the persuasive entrepreneur. I was a young attorney during that time, representing several of the companies the author examines, and I am grateful for the insights he offers here. He has captured the remarkable excitement of that period. Of course there were episodes of investor loss, but also, for some, the creation of extraordinary wealth. Author Donald Hall lived through the times as well, and has found many sources whose memories bring life to that history.

The success in raising public money spawned the formation of additional capital from small business investment companies, and encouraged bank lending to small, unproven companies. Directors from larger and older publicly owned firms joined small company boards and introduced improved governance, quarterly reports, better methods of evaluating company officers and board performance, and the formation of compensation committees. The combination of these factors brought into being, or strengthened, young companies that today are major sources of employment, product creativity, and exemplary corporate governance. We can be thankful for the extraordinary developments of that earlier period.

The question now is whether new business ideas have dried up, or whether our standards of risk have made us unduly cautious in the commitment of capital. I ask you to read on and learn the spirit of a time, some fifty years ago, and decide for yourself whether replication today would be healthy or foolish.

– Thomas E. Holloran
Senior Distinguished Fellow,
Thomas Holloran Center for Ethical Leadership in the Professions, University of St. Thomas School of Law
Former president, Medtronic

Preface

In 1961, I was enrolled in graduate school at the University of Minnesota and working part-time in the trust department of the First National Bank in downtown Minneapolis. Finance was my major. Before receiving my degree I was required to write a thesis, which I found daunting; I could think of no worthy subject. My faculty advisor suggested I talk to Merrill Cohen, then head of the J. M. Dain brokerage firm in the Minneapolis financial district. I was pleased about the opportunity to get assistance from someone directly involved in the securities industry.

"I think this market we're in, right here, would make a good subject," Mr. Cohen said. "It's unusual. I can't think of too many others like it. Maybe Salt Lake City in the early 1950s. I'll be curious to see how it all ends. The valuations are crazy." Then he emphasized his point. "It would be a good subject for a thesis. Control Data started it all."

I had heard about Control Data through ordinary conversation and knew it was a stock market sensation, but that's about all I knew, so I paid close attention to everything Mr. Cohen said, and asked a question or two. He finished the meeting with a gentle command. "I consult with students like you, give them my best ideas, and never hear back. If they act on them, I never know about it. Get back to me sometime, and let me know what you've done." I nodded yes.

I liked the idea and felt energized, but also inadequate. How could I determine the essence of a market I had just happened into? Where would I find the necessary information? And how could I capture the feel of something I had not yet felt? I had no experience in such things. Nevertheless, I learned what I could about the local market and tried to write a few opening paragraphs. But the writing stopped there. What I had written seemed forced.

Soon I married and began working full-time. The need to complete my degree faded into the background, but never completely, as it continued to fester in that dark corner of the mind called unfinished business. The merit of Mr. Cohen's suggestion never left me. And as events unfolded in the local market, I came to realize that he was right; the subject was unusual and it had true financial significance.

Upon reaching retirement, I began to reconsider the idea of writing this history. Almost fifty years had passed since Merrill Cohen made his suggestion, certainly enough time to reveal the meaning of it all. I had worked as a security analyst for the First National Bank of Minneapolis, and then for thirteen years at Control Data Corporation. I was lucky enough to have formed an informal acquaintance with most of the company founders, since our offices were on the same floor. I also had the auxiliary assignment of driving important visitors to company plants for interviews with management. This gave me an unusual opportunity to observe and listen to Seymour Cray at his plant in the woods near Chippewa Falls, Wisconsin—something most employees were unable to experience. After leaving Control Data, I worked for twenty years in the local securities industry. During that time I accumulated pertinent financial information and gained experience in interpreting stock market behavior. Re-energized, I set about researching this book. Mr. Cohen, I am now ready to report back and thank you for your suggestion. I hope you can hear me.

The principal purpose of this book is to describe an unusual period in stock market history, and beyond that, to illustrate how the success of a single company, Control Data, made it much easier for other "start-up" companies to secure financing in pursuit of their own dreams. Many of the individuals who invested in these stock offerings were hoping to capitalize on a success similar to Control Data's. Indeed, the influence of the Control Data story on other young businesses is an important part of its enduring legacy. The medical device industry thriving in the Upper Midwest today can be considered an outgrowth of that earlier investment environment.

But it takes more than financing to build a strong company. They spring to life under the leadership of extraordinary individuals. A history of the times could not be written without describing the abilities of William Norris, Seymour Cray, Arnold Ryden, Willis (Bill) Drake, Zinon (Chris) Possis, Earl Bakken, and Manny Villafana, all of whom founded companies—some more than one.

The story I'm telling here is not one of universal success. Many promising ventures had a short life. Tonka Toys at one point produced almost as many (toy) vehicles as General Motors; its factory is now owned and run by other businesses. Possis Machine invented unusual devices—first industrial and then medical; it's been swallowed up by another firm. Some companies disappeared quietly without achieving noteworthy success while others—Medtronic and St. Jude Medical, for example—continue to thrive in a local medical device industry that by one estimate now includes some 350 companies. All were financed somehow. All needed technical talent. Control Data provided the example, inspired the investment, and in some cases supplied the people to make them go.

There are lessons for the individual investor in these pages, too. Descriptions of what worked and what didn't in a given case should help readers sort out elements essential for success.

Recently, the Kauffman Foundation issued a report ranking Minnesota dead last among the fifty states in entrepreneurial activity. This was a shocking revelation to those of us who remember the nurturing environment of earlier times and the prosperity that followed. The best characteristics of those times should be studied, celebrated, and brought to life again. I hope this book contributes to such a revival.

– Donald M. Hall
February 2014

Generation of Wealth

There is a tide in the affairs of men.
Which, taken at the flood, leads on to fortune...
And we must take the current when it serves,
Or lose our ventures.

– William Shakespeare

1

The Birth of a Champion

There was "a seething resentful atmosphere" among the employees at the St. Paul Univac division in early 1957. Their work was not being funded adequately by the parent company, Sperry Rand, and they found themselves competing with another computer development operation that Sperry owned in Philadelphia. The two organizations looked upon one other with distrust, if not outright contempt. They had both been acquired in the early 1950s for their pioneering work in computer technology, and once acquired, their individual responsibilities were never clearly defined, or, if defined, were never fully accepted. When Lawrence Radiation Lab in Livermore, California, issued its specifications for a scientific computer, both organizations responded and each insisted on having primary authority for the project. Both organizations also thought they had the better technology to penetrate commercial markets.

The Philadelphia branch of the firm had produced a general purpose computer the size of a garage, whose blinking lights suggested immense intellectual power. On election night, 1952, it received national exposure over the new medium of television when early returns allowed it to forecast that Eisenhower would become the next president of the United States in a landslide victory, even though most polls at the time showed him trailing. The broadcasters at CBS distrusted the startling prediction and were reluctant to mention it on the air. But the actual count of ballots brought validation, as Dwight Eisenhower overwhelmed Adlai Stevenson, winning thirty-nine of the forty-eight states and 55 percent of the popular vote. For the machine and its builders, it was a moment of triumph. But such moments were not the norm. The computer consisted of thousands of vacuum tubes and electrical connections, any of which could fail unexpectedly, giving engineers the tedious job of locating and fixing the problem while customers waited in frustration. Normal maintenance on the machines took 20 percent of available time.

The St. Paul employees, on the other hand, built computers for the U.S. Navy that were required to operate under unusual environmental conditions such as rolling seas and salt air. Their machines were reliable. The engineers responsible felt their work was clearly superior to that being produced by the Philadelphia group, though it was difficult to get that point across because many of the technical breakthroughs they achieved were classified "top secret." To discuss them even with their own corporate management would be a felony.

At corporate headquarters in Norwalk, Connecticut, executives gave neither organization proper attention, choosing instead to focus on electric razors, typewriters, and other company products they knew would make money. What the computer operations lacked was leadership that understood and shared their vision, and trusted their abilities enough to fund their work. Some Univac employees, primarily engineers, considered themselves so poorly managed they began to look for other work.

Arnold Ryden had worked among the St. Paul group in earlier days as company treasurer. Now he was a management consultant, specializing in finance. He was aware of the widespread discontent among Univac engineers, and he worried that individual defections would, in time, deplete the pool of exceptional talent. Quietly, he began conversations with key employees about forming a new business. They listened with interest but wondered how the required money might be raised. In a separate conversation some of the employees counted up their personal investment possibilities and came up with a cumulative total of just $15,000. They also doubted that brokerage firms would be willing to help fund a handful of individuals needing significant cash for a vague business idea, since adults were only one generation removed from the Great Depression and its lessons about investment risk. But Arnold Ryden saw it differently. He was a businessman, not an engineer; he knew balance sheets, cash flow, and financial possibilities. "We can sell the stock ourselves," he said, and he outlined plans on a sheet of paper, confident it would work.

Bill Drake

Willis (Bill) Drake became an early ally. An engineer by education, Drake had worked at a Univac customer installation and suffered agonizing difficulties as various pieces of input/output equipment promised by the Philadelphia operation failed to arrive at the customer site. Upon investigation, he learned that the people involved in designing the equipment never considered their work finished and redesigned it repeatedly. The customer, meanwhile, tired of delays, threatened to return the computer and cancel the project. Drake implored corporate management in Connecticut to get involved. It took him nearly two years to finally bring about a satisfactory conclusion. Now he was back in the Twin Cities, a sympathetic listener to Ryden's idea for a new company. Both Ryden and Drake knew that their principal objective would be to gain the support of William Norris, the leader of the St. Paul employees, but they also knew that in order to succeed, the new company would need a core group of talented engineers, so they engaged other candidates in private discussion.

In addition to Norris, Frank Mullaney was crucial, as he managed the St. Paul military systems group. In February, Ryden invited Mullaney to the Minneapolis Athletic Club to discuss his plans. Mullaney listened, gave the proposal serious thought, but declined to get involved. Nevertheless, he told his boss, Bill Norris, about the conversation, and was perplexed when Norris showed little reaction. Norris could sometimes be that way, Mullaney told to himself, and considered the matter finished.

Frank Mullaney

But Norris, too, had been frustrated over the constraints placed on the business, and was bitter at seeing a large market opportunity escape his grasp. As leader of the division, however, he felt he should fulfill his management responsibilities and keep his personal thoughts to himself. When employees hinted that they would like him to start a new company, he let their remarks pass.

But in April 1957, Norris's attitude changed. The Univac division had been reorganized, and he had been demoted. More than that, he felt he had been encouraged to leave. His employees noticed, too, as did Ryden, the consultant, who described the changes by saying Norris had been "emasculated." Norris now felt free to express himself fully and to explore other alternatives.

He invited his trusted subordinate, Frank Mullaney, to lunch, and this time told Mullaney that he had been aware of the effort to start a new company—even talked about it to his family as early as January—but had kept his thoughts to himself. He expressed his disgust with the recent organizational changes at Univac and said, "I've had it." He wanted to know if Mullaney was still interested in such a venture. Mullaney's eyes brightened. Yes, he was interested, knowing that Norris would be crucial to the success of the company.

Word spread quickly within the St. Paul Univac operation about the possibility of forming a new company, and employees began to call Norris with questions. He eventually decided to clear the air by convening a small gathering at his house; he was flabbergasted at the number of people that came. His neighbor, he later said, thought there was a funeral. In a sense there was, but it was overshadowed by the possibility of new birth.

Norris committed himself to lead the informal group now ready to follow him. Describing his fundamental difference with the management of Sperry Rand, he said later, "They just weren't willing to take the risk. If we'd had financing we might have become the IBM of the industry. We were much more advanced."

Ryden and Drake drafted a prospectus for the new company on Drake's kitchen table. Together with Byron Smith, they had experimented with an appropriate name by writing individual words in two columns and considering the various combinations. They arrived at *Control* and *Data* (pronounced *day-ta*). It seemed simple at the time, even mundane, but they accepted it as the best choice. When Norris heard the name, he was unimpressed, but he, too, failed to come up with a better alternative. Nor could anyone else. The name *did* have a lyrical inflection, and it conveyed meaning, even some originality. In time, it came to sparkle with a connotation of technical success, an escalating stock, and almost impossibly quick wealth. It came to captivate many in the Twin Cities area, and even on Wall Street and the nation's other capital markets.

Uncomfortable organizing a new company while working at Univac, Drake decide to resign and devote all his time to the new effort. He was paid a modest salary with funds put up by Norris and Ryden, and he rented space at the McGill Building at 501 Park Avenue near downtown Minneapolis. Though he had no experience raising capital, he was young, personable, and eager to address any task. His initial responsibility was to deliver a draft of the prospectus to the Minnesota securities commissioner for review and approval

Selling stock to the public without an underwriter was an unusual approach, the commissioner noted, but he found no technical objection. Still, he was troubled that the company had no product, no customers, no plant, no money, and almost no employees. Although they seemed like nice people, he said, he doubted they would succeed. Still, he wanted to understand how they thought otherwise, and asked Drake to "explain this one more time." To help ease his concerns, he requested a list of prospective investors. More meetings followed. Finally, he insisted that each investor sign a form stating there were no assets in the business, and there was no way to value it.

On July 26, Norris resigned from Univac. The prospectus for the new company, published three days later, showed that the company had been organized on July 8, with Norris, Arnold Ryden, and Fremont Fletcher, an attorney, the original shareholders. The price of the stock was set at $1 a share—cheap enough to appeal to anyone. Surprisingly, sports writer Sid Hartman was first to provide public exposure for the new company when he mentioned it with a leading question in his column in the *Minneapolis Tribune* on July 30, 1957, under a sub-heading called Off-the-Sports-Beat. "Have several former Univac people resigned the St. Paul Sperry Rand branch to form a firm called Control Data that will deal in electronic research?" he asked, and left it at that, a portentous question. The *St. Paul*

Pioneer Press mentioned the new company and its officers in an article dated August 15, 1957, and the *Minneapolis Star* cited it briefly in the "Business and Markets" section on August 26, 1957. Neither article mentioned stock being sold.

Norris made a presentation to the Twin Cities Society of Security Analysts at the Town and Country Club located along the Mississippi River on St. Paul's western edge, but also easily accessible from Minneapolis via the Lake Street bridge. Attendees came from both cities, probably twenty in all. This was a group not normally interested in speculative new issues—they were analysts and there was little in the proposed new issue to analyze—but they listened with interest, and upon returning to work likely suggested the stock as consideration for some of their more venturesome clients. Jim Miles, a Univac employee intent on joining the new company, together with his wife, Laura, held a dinner party for friends at their home in southwest Minneapolis. When the meal was over, the men retired to the garage, where Jim invited them to buy shares in the new company. Norris induced a classmate from college who had become a physician to invest.

William Norris

Wheelock Whitney, a prominent local businessman, philanthropist, and sometime politician, was working in investments at the time; he attended a dinner party and was urged to invest by his friend Don McNeely. The company "needs your Midas touch," McNeely said, and he suggested a commitment of $25,000. Whitney used normal business prudence and consulted with a few individuals knowledgeable about computers. They responded that International Business Machines had a stranglehold on the industry, and that other big companies with huge resources were moving into the field. The new company, he was told, "didn't have a ghost of a chance." When he declined to invest, McNeely told him he was making a mistake. But Whitney was not alone in his decision. Warren Buffett, who was related to Norris by marriage and later became an investing legend, declined also, feeling he did not have the technical background necessary to evaluate the stock.

Meanwhile, Drake held informal meetings and lunches with whomever he could, hoping to help raise money. After many such meetings he finally got an order from a husband and wife for a few thousand shares, but it was tedious work, and slow; Drake and Ryden eventually decided it would be better to hold house parties where they could address larger groups of people. They invited a dozen or so to their first effort. After hearing the proposal and having various questions answered, an individual offered to buy fifteen or twenty thousand shares. This emboldened the others, and four or five more signed up. Enthused by his success, Drake held a second party, inviting another dozen people. This time thirty showed up. So Drake held a third party, inviting another dozen prospects to his home in Edina. Word spread quickly among the workers at Univac and that night cars were parked all over Drake's neighborhood. Forty or fifty people had shown up, eager to hear the story and also to capitalize on the new opportunity. Drake reminded everyone that the venture was speculative. He probably also told his listeners that Norris was putting personal money into the new company at the same price as initial investors, a fair deal for everyone. The prospectus, only eight pages long, added to investor knowledge by saying the company would initially engage in:

"...the sale of services in research and development engineering, largely for military end-use. Research and development services will also be made available to companies engaged in commercial electronic data processing... an industry in the early stage of development. It is anticipated that the Corporation's commercial activities will expand in the areas of input and output equipment, data recorders, data translators and converters and a wide variety of other peripheral equipment for electronic computers."

The prospectus specifically stated that the company did "not plan to compete directly with the giants of the industry, such as IBM, Sperry Rand, General Electric, Burroughs, etc, except for certain carefully selected military programs." Instead, it said, "plans are to supplement these major companies as an important subcontractor and as the developers of selected components and equipment which may be used with the computer systems and instrumentation of the major companies." It also said the company was negotiating for "10,000 square feet of temporary space for six to nine months and for 40,000 square feet of space with expansion possibilities up to a total of 200,000 square feet in early 1958," and that "negotiations were being conducted with several small engineering and production companies which may result in the acquisition of one or more companies." Norris would receive an annual salary of $20,000 and Ryden $16,000, with higher remuneration considered after the company reached "substantial operations."

David Lundstrom, who would later work for the company, was present for Drake's third meeting; he was intrigued, but skeptical. He bought 200 shares at the offering price of $1 a share. Later he recounted in a splendid book of recollections that if he had bought 1,250 shares, an amount he could easily have afforded at the time, he would have become a millionaire in eleven years. However, had he asked for those additional shares, he might not have received them. Another Univac engineer, who prefers to remain anonymous, was also at one of those meetings and says that he requested 8,000 shares but was given only 2,000. That investor also knew of someone whose request was turned down completely, probably because the subscriber wasn't a likely future employee or a close associate. But one of Drake's neighbors, curious about the commotion next door, came over, listened to the story, signed up, and was given stock. His intuitive decision in time made him a millionaire and changed his life.

A month had passed, and the new company now had subscriptions for $1.2 million, double the amount offered in the prospectus. Drake brought the necessary forms to the state securities commissioner and placed them on his desk. The commissioner smiled in acknowledgement. Drake recalled later that people called the commissioner and others from around the country and asked, "Is this true?"

The experiment in having employees sell stock directly to the public rather than through an underwriting broker succeeded quite easily. The company accepted $615,000 from approximately three hundred investors. Norris himself mortgaged nearly all of his assets and invested $75,000, an impressive sum, since $75,000 at the time could have easily bought three quality houses for cash. "If we lose it all, we'll just move ourselves and six children back to the farm in Nebraska," his wife, Jane, said, and she invested some of her own money as well. The new company began business in September 1957. Those unable to get stock at the initial offering were free to buy once trading began in the public market.

The national trade newspaper *Electronic News* had been checking sources, trying to pick up information on activities in the Twin Cities; they knew that a new company was being formed, and on August 5, 1957, reported: "A firm called Control Data is in the process of opening. Trade reports indicate William C. Norris will be instrumental in operation of the firm."

Two weeks later, they commented on the sale of Control Data stock, "reportedly sold by Mr. Ryden." And on October 20, the paper described Norris as "a quiet, intense man, afflicted with what associates call a scientific restlessness." Engineers all across the country read this newspaper, but the articles, while very early with the news and generally accurate, were brief and didn't attract much attention.

William Norris and his twin sister, Willa, were born on July 14, 1911, on a farm of a thousand acres in south-central Nebraska, very close to the Kansas border. Norris's grandfather had homesteaded the farm, at the time not far from Indian territory, in 1872, and had picked up additional acreage from time to time as it became available. White settlement brought glistening cornfields to the heavy lowland soil in that stretch of Nebraska and fenced pastures to the rolling, higher ground.

William and Willa Norris

Young William was required to milk the family's five or six dairy cows and help feed the hogs and beef cattle. "It was a hell of a nice life," he remembered. One arid summer, the Republican River bordering the property dropped so low that cattle waded across to feed in pastures on the other side. When heavy rains flooded the riverbanks, the boy helped his father bring the cattle home. Since the nearest bridge meant a drive of six or seven miles, they made the cattle swim the river, compelling them on horseback, a task that required crossing the river back and forth several times. "It was exciting," he remembered, "as I was not given to daring physical feats."

Willa Cather, the American author who wrote about pioneer life on the Great Plains, lived in the nearby community of Red Cloud. According to some accounts, she was a friend of Norris's mother, who named William's twin, Willa, after her; however, one of William's daughters says there was only passing acquaintance between the two families and that her grandmother named both William and his twin after their father, also named William. The Norris twins had an older sister named Katherine, who married Fred Buffett and became an aunt to investor Warren Buffett, but there was little social interaction between the two extended families.

Like most rural homesteaders, Norris attended the first eight grades at a one-room school house which he often reached on horseback. To get to high school, eight miles away in Red Cloud, he and Willa, then thirteen, drove the family Model T. There were likely few other cars on the road and no license requirements. Red Cloud, named after a Lakota chief, received rail service in 1879 and had grown to be a community of 1,800 people by 1920. William played in the line on the high school football team. Willa got better grades than her brother but only "because he didn't apply himself," she said later in an interview. "But in physics, his favorite subject, he knew more than the teacher." Both the twins were smart. Willa became a professor at Michigan State University. (Later in life, due to her warm personality, Willa was appointed Vice President of Mingling for Control Data Corporation—a post she thoroughly enjoyed.)

William loved to fish, hunt, and trap, and he collected bounties on the skins, including skunks and badgers. In later years, he sometimes used an earthy metaphor in business discussions, advising colleagues not to "get in a pissing contest with a skunk."

Like most future engineers, William tinkered with electronics, ham radios, and model airplanes as a youth, and he liked to read *Popular Mechanics* magazine. He later pursued those interests formally in Lincoln, earning college expenses as a radio repairman and graduating from the University of Nebraska in 1932 with a degree in electrical engineering. "He was a real good student," his roommate recalled. "He had a real determination to succeed, but had time for a little social life on the weekends, too." Shortly before graduation, however, William's father died of a heart attack, leaving William the sole male in the family. Upon receiving his diploma, he went back to work on the family farm.

The young William Norris

The Great Depression was underway by 1932, but more importantly, on the Great Plains drought had gripped the land, limiting the production of grain and hay and stirring up dust storms and swarms of grasshoppers. Without an adequate crop, Norris had to consider sending the cattle to market at disastrously low prices or somehow keep them nourished until conditions improved. As a youth he had noticed the cattle occasionally eating Russian thistle before it turned brown, broke loose, and blew about as tumbleweed. During the drought that summer, the thistle grew abundantly, while the small grain never matured, and the corn withered. Norris chose to harvest the green thistle and store it for winter feed. Other farmers were reluctant to do so; they doubted whether the cattle would eat it, or receive sufficient nourishment if they did. Norris, on the other hand, was so sure of his gamble that he bought more cattle and enhanced the thistle with feed supplement. The cattle were able to remain healthy over the winter and were strong enough to calve in the spring when the pastures turned green from early rain. That lesson "greatly added to my sense of self confidence," he recounted in later years, and it gave him a deep trust in his own judgment and a willingness to ignore the uncertain doubts of others. He told the thistle weed story more than once when asked about his formative years.

Between 1932 and 1934, the nation's unemployment rate exceeded 20 percent. Nevertheless, two years after college Norris received two job offers from Westinghouse, the big electrical products company: one in engineering, and the other selling x-ray machines. He preferred the engineering position but took the sales job because it paid twice as much, and he ended up sitting in a lot of doctors' waiting rooms reading magazines. Nevertheless, in his own words, he "became a pretty damn good salesman." It must have taken special effort, for Norris didn't have a salesman's natural outgoing personality, but he did have the power of intellectual persuasion, considerable self-confidence, and a disarming, bashful grin. He had a controlled, somewhat monotone voice and the lean, tanned look of a man from the country who might have been wearing a feed cap.

In 1940, the U.S. Navy Bureau of Ordnance was hiring engineers in Washington D.C. Seeing a notice for the job in the post office, Norris applied and was hired. He moved to Washington and was placed in a bullpen with other engineers checking and revising blueprints for antiaircraft guns on naval ships. Although he said he enjoyed the work, it soon became tedious, and he started looking for another opportunity. When war broke out, he got his chance.

Thirty years old and not yet married, Norris applied for and received a commission as lieutenant junior grade in the

U.S. Navy Reserve. Although he never understood quite how it happened, the Navy assigned him to a unit engaged in super-secret cryptological warfare, also called code-breaking. German and Japanese radio transmissions were intercepted, analyzed, and decoded—a task that involved recording and processing enormous quantities of data. Norris and his team had to find the best means for quickly processing this vast and continually updated flow of information, looking for patterns of frequency and repetition of certain letters or characters. To meet this need they designed equipment to reduce the information to binary simplicity and analyze it electronically. Ever the inquiring mind, Norris, while in Washington, also took advanced mathematics and engineering courses at George Washington University. In addition he developed an affection for an office secretary, Jane Malley, six years his junior. It was a busy time.

Jane Malley had grown up in Parkersburg, West Virginia, received a college degree in education from the University of Cincinnati, and graduated and received a commission in the first class of WAVE (Navy auxiliary corps) officers from Smith College in Massachusetts before being assigned to the nation's capital. In September 1944, she and William were married in Washington D.C., wearing their uniforms. The union would last sixty-one years and produce eight children.

When the war ended, most military activities were deactivated, but the Navy felt it should maintain its intelligence processing capability, as mistrust in the Soviet Union was already a major concern, and the group of code-breaking specialists the Navy had rounded up and nurtured represented an unusual national resource. Norris and some of the others considered forming a private business. Secretary of the Navy James Forrestal approved the idea of the group staying together, with the implication that it would be able to receive government contracts.

Principals in the new company would include Lt.

Commander William Norris, Commander Howard Engstrom, a mathematician, and Captain Ralph Meader from the Naval Computing Machine Lab in Dayton, Ohio. John Parker, an Annapolis graduate and a successful investment banker, put together the ownership agreement. Employees would get one-half ownership, or 100,000 shares for 10 cents apiece; investors would get 100,000 shares also for 10 cents apiece but would underwrite a $200,000 bank line of credit. Total equity in the company was $20,000. Investors unable to come up with the required investment could borrow at the First National Bank of St. Paul, where Parker had agreed to sign for them. The company was incorporated in January 1946 under the name Engineering Research Associates (ERA).

Parker, as president, was based in Washington D.C., as was Engstrom, working with government scientists, while Meader was given responsibility for operations, and Norris, engineering. Within a year, however, Norris would be put in charge of marketing. As leader of a technology company doing top secret work, Parker, an investment banker, acknowledged his unusual role by saying it was "like taking over the symphony orchestra without knowing a note." Five times Parker had tried to strike a financial arrangement with various firms, including National Cash Register (which managed the Naval Computing Machine Lab in Dayton, Ohio), Raytheon, American Airlines (probably to develop a reservation system or an air traffic control system), without success. He then enlisted private individuals.

Although a Washington D.C. resident, Parker was president of an aeronautical company whose St. Paul plant at 1902 Minnehaha Avenue, in the Midway area, produced gliders to transport army troops and related cargo. He also served on the board of directors of St. Paul–based Northwest Airlines. With the war over, the aeronautical plant stood empty. He chose to locate the new company in that unlikely northern location, a cavernous 140,000-square-foot building illuminated by skylights and home to noisy sparrows and pigeons and their inconsiderate droppings. It resembled a hangar and was just as cold and drafty—during the winter months workers found it necessary to wear mittens—but it did have a concrete floor and a balcony with offices. Heating units were added and fluorescent lights were hung on chains from the ceiling. Additional offices were constructed with sheetrock partitions, and a contingent of thirty to forty employees associated with the earlier code-breaking work in Washington D.C. moved to St. Paul, joined by a few workers from the Naval Computing Machine Lab in Dayton, Ohio. By the end of its first year, the company generated revenue of $1.5 million, primarily from cost-plus-fixed-fee contracts for the U.S. Navy, and recorded a profit of $34,000.

The entrance of the ERA plant

The Minneapolis/St. Paul area may have seemed an unlikely location for the new company, but there was already technical talent in area. Minneapolis Honeywell Regulator Co. did government work, and so did General Mills. Yes, General Mills, the cereal company. It had started a mechanical division in the 1930s to make processing equipment for internal production needs, and one thing led to another. During the war the company produced naval gun sights, radar antennae, depth charges, Air Force bomb sights, and high altitude plastic balloons. Eventually a computer capability evolved. "Old line" companies such as Buckbee-Mears and the Bureau of Engraving began making printed circuit boards and etching electronic components. Minnesota Mining and Manufacturing was very much a technical

John Parker

innovator, but mostly in commercial products. There were also some smaller technical companies: Automatic Control, Control Corp., Mack Engineering, Minnesota Engineering, Precision Inc., and four separate hearing aid companies. Feeding these activities was the University of Minnesota, graduating a new class of engineers each year who were looking for work.

In 1950, Maeder wanted Parker to sell ERA to IBM; Parker chose instead to buy Maeder's stock for $3 a share, a thirty-fold increase from the initial subscription price. Maeder resigned, and Parker put Norris in charge of operations. That year, the company delivered a general purpose computer that could be "programmed," an innovation at the time. But the company's accomplishments were quiet ones. E. Douglas Larson, an employee, remembers that the company shipped a computer called ATLAS II to the Navy—a machine so secret the government didn't even reveal its existence until twenty-four years later. Employees working on one part of the machine had no knowledge of other parts or the final product. Virtually all of the company's computer business was done for the government.

ERA had never been sufficiently capitalized, however, and a chronic shortage of cash, along with heavy borrowing, strained operations. "Frantic finance," Bill Drake called it, remembering that somebody ran to the bank every Friday to meet payroll. So, in 1951, Parker decided to sell the company to Remington Rand for 73,000 shares of common stock, or approximately $1.7 million. The stock was listed on the New York Stock Exchange and represented $8.50 a share, or 85 times the initial cash investment undertaken only five years earlier. Various sources give slightly different dollar amounts, but the number of shares is not in dispute and the different dollar figures could represent market quotations on different days around that time. The price was reportedly set by multiplying the number of engineers, 340, by $5000, a somewhat unusual method of valuation, but justified as the price it would take to rebuild the staff. Net worth in the company was just $150,000, and debt, twice that amount. It was the unique talent that held value. Interestingly, employees other than engineers were not factored into the equation. There were approximately a thousand total employees. The acquisition of ERA, someone quipped, was one of the few times human beings had been sold that way since the abolition of slavery. In any event, the sale resulted in a big investment success for those few who held stock, including Norris, but to most of the employees it was like "a bucket of cold water." They wanted control of their own activity.

Norris agreed to go along with the decision to sell, but not before insisting that "it was the wrong thing to do," since he was convinced the company had a bright future on its own. His disinterest in quick financial success reflected his personal attitude in general, for in the years that followed, when financially comfortable, he persisted in wearing suits until the fabric glistened, drove the same Oldsmobile Cutlass for many years, and

Workers at Engineering Research Associates (ERA)

lived a quarter-century in the same house. He had no interest in the outward symbols of success. For entertainment he enjoyed fishing and attending Minnesota Twins or Vikings ball games. But he always had high ambitions, it seemed, like proving to some inner critic, or maybe outer ones, that he knew what he was doing.

Once the acquisition was complete, Norris found himself travelling to Remington Rand headquarters in Norwalk, Connecticut, for conferences with General Douglas MacArthur, the World War II hero of the Pacific, then serving as chairman of the company. MacArthur enthralled attendees with stories about the war, but didn't have much to do with day-to-day operations. His principal value was in cultivating customers. Like the others, Norris enjoyed his association with the former war leader but got little in the way of business direction from him.

Four years after buying ERA, Remington Rand was itself acquired by Sperry Corporation, and the combination was renamed Sperry Rand. Since Sperry had a strong technical orientation, the St. Paul group thought they would get better support for their work. Norris, for a time, was put in charge of all computer development activities, including the group in Philadelphia and a lab in Connecticut. Combined, the activities came to be known as the Univac Division. But things did not go well. Norris had to justify the need for development funds repeatedly, and he had difficulty moderating the rivalry between the various operations. Meanwhile, he watched his principal outside competitor, IBM, become very successful. Its stock had gone up tenfold since the end of the war as the company capitalized on business data processing needs. Norris felt disheartened by the lack of the support he received from corporate headquarters. Then his division was reorganized.

"When they started breaking up the Univac Division, part by part, limb by limb, I left," he said. He was forty-six years old, experienced, and determined to prove his convictions. Once Control Data was formed, Norris said: "My decision to leave Sperry Rand was based primarily on my utter distaste for the whole thing," and he later recalled, "I didn't really have a well-thought out business plan in mind. I just wanted to get the hell out of there, and felt the field was large, it was expanding, and we would be able to develop a product as we got the company started."

As stock was being sold to the Univac employees in Drake's living room, many likely knew of the spectacular eighty-five-fold increase in value that the earlier ERA Corporation stock had experienced in five short years. They may have known that Norris had only reluctantly acquiesced in that sale, concerned that a huge opportunity would get lost in the corporate entanglement that would accompany it. Now that opportunity could once again be pursued. Robert Perkins, an early ERA engineer who would join the new company, said, "Control Data became the second ERA."

With money in the bank, the new company began to hire a core group of employees, almost all of them electrical engineers. It had four employees at the beginning of September 1957 and twelve by the end of the month, including Norris as president; Frank Mullaney, Bill Keye, Bob Kisch, Howard Sheckels, Bob Perkins, and Seymour Cray, all technical people from Univac; Bill Drake, Hank Forrest, and Jim Miles, all of whom specialized in technical sales, also from Univac; plus Arnold Ryden, whose background was finance; and Lucille Walker, secretary. There was no need to recruit talent; many were eager to come. Employee number thirteen, Univac engineer Edward (Pete) Zimmer, was soon followed by Jim Thornton, also carrying a slide rule. Both would have a significant impact on subsequent work. This was a smart group of people. Bob Kisch and Frank Mullaney discovered that their birthdays were very close and that they were born in the same year, but Kisch knew that he had advanced a grade in school. One day he was puzzled to discover that Frank Mullaney's class reunion was ahead of his. Mullaney, he learned, had skipped two grades, and in fact had started college at the age of sixteen. Mullaney, a bespectacled, six-and-a-half-foot man of pleasant demeanor, was not inclined to boast of his accomplishments.

Two months after the public offering, as hinted in the prospectus, the company acquired Cedar Engineering for a combination of cash and stock. Located in the nearby suburb of St. Louis Park, Cedar Engineering made instruments and controls, and was generating revenue of approximately $2 million per year. This quickly added to Control Data's credibility in pursuing production contracts, as Cedar Engineering had "clean room" manufacturing facilities and 178 employees. Control Data could now say it was a real business, not just a group of engineers willing to design products.

Meanwhile, one of the Control Data engineers, Seymour

Cray, designed a modular circuit configuration that could be used in anything digital, including high-speed computers. "It was original and it was fast," Norris explained later. "We didn't really know what we were going to do [when we started]—at least, I didn't—but when Seymour came up with the idea of building a very big computer and the means for doing it, he gave us something to tie into. I kind of felt like shouting, 'Eureka!'" Norris then directed company resources to support Cray and his small team of designers. With that, Control Data suddenly embarked on a program to build the world's most advanced computer, and ultimately to compete with the other major companies in the industry. The *Minneapolis Star* on November 11, 1957, quoted Norris as saying, "We're aiming at sales of $25 million a year within five years." He cited the company's ability to make specialty printed circuit boards, i.e. "building blocks," with transistors, which were more efficient and cheaper than vacuum tubes. *Electronic News* picked up a similar quote on December 2, 1957. Investors had reason to be optimistic.

Cray, then thirty-two years of age, was already recognized as an exceptionally brilliant man, even a genius. He could easily intimidate and sometimes humiliate a colleague not up to his intellectual capacities, and seldom considered the need to behave otherwise. But that was not his intention; his mind simply worked on abstract complexities, not social niceties. Pete Zimmer, a fellow engineer, remembers a meeting when the technically accomplished Frank Mullaney joined the group. "I don't think you'll understand what we'll be talking about, Frank," Seymour said to his boss. Mullaney, with employee-centered reserve, smiled and let the comment pass.

Around the office, Seymour was introverted and unassuming. Walking down a hallway, he looked like a neatly dressed, average-sized man, minding his own business. If you caught his eye, he would smile as though enjoying something interior and mysterious, but mostly he didn't make eye contact. Generally arriving at 11 in the morning, he often worked beyond midnight, preferring the quiet of that time. He kept a clean desk, the work environment reflecting the clarity and order of his mind. When he found the telephone a distraction, he had it removed. He also requested that the loud-speaker system be disconnected. Once he had the right conditions, colleagues began to notice his remarkable ability to concentrate.

Associates, even customers and outsiders, came to call Cray by his first name only, Seymour, the singular identification of a person everyone recognized. His last name, however, evolved into a sardonic verb form. When Seymour Cray reviewed their work, fellow employees sometimes came away slightly shaken, and said they had been "scrayed." Seymour communicated very directly, Pete Zimmer said. His technical insights could be brusque and withering, but he was an otherwise pleasant man who accepted the fact that he made an occasional mistake himself. Colleague D. (Sam) Slais observed that when such a mistake did occur, Seymour would acknowledge it, correct his course, and proceed in a new direction. "He was a very nice guy," said Mike Schumacher, another colleague. His outsized fame arose not from any dramatic behavior on his part, but from his quirks and an awestruck appreciation of his work.

Seymour Cray was born in Chippewa Falls, Wisconsin, on September 28, 1925, five years ahead of his only sibling, Carol. The name Cray was of English origin. Both mother and father encouraged his technical interests, and as a youth he performed experiments with his chemistry set and once assembled a functioning short-wave radio in the basement. By age ten he had built an automatic telegraph machine. Eventually his tinkering reached the bedroom, which he wired so that an alarm would go off whenever someone touched the doorknob. Seymour's mother, a homemaker, accepted her son's inclinations and simply gave up cleaning the room. His father was the city engineer in the community of twelve thousand. In him, citizens had benefit of a single authority for plumbing inspector, electrical inspector, building inspector, and whatever else was necessary. Perhaps explaining his own introversion, Seymour once noted that his father relied on his wife for social interaction.

In high school, when the physics teacher was absent, Seymour was said to have filled in for him. "I was one of those nerds before the name was popular," he admitted. "I spent all of my time in the electrical engineering laboratory and not enough time socializing." Seymour did sing in the boys' glee club. In spite of his intelligence, it's interesting to note that he ranked only sixth in his graduating class by grade-point average, though he did receive the school science award.

Upon graduation in 1943, Seymour was drafted into the U.S. Army and placed in an infantry communications platoon, serving first in Europe and then in the Philippines. Those assignments may have been the first and last times he submitted to any real outside authority over his work. When the war ended, he returned home and married Verene Voll, the daughter of a local minister. For a short time, he attended the University of Wisconsin, and then he transferred to the University of Minnesota, earning an electrical engineering degree in 1949 and a master's degree in applied mathematics at the turn of the year in 1950–51. Engineering professor Paul A. Cartwright of the University of Minnesota described Seymour as the brightest student he ever taught. "When I learned where he was working," Cartwright added ruefully, "I should have bought the stock."

E. Doug Larson, a physicist and fellow employee, had another observation. "Seymour was an excellent chess player," he said, "giving away pawns freely, and then suddenly announcing a surprising, strategic checkmate. And he played fast." Seymour also played a superior game of poker, both Larson and Zimmer reported.

In 1951, Seymour began working at ERA. Within weeks of joining the company he began offering suggestions on how work should be done, and other employees, accepting his superior understanding, proceeded accordingly. By power of intellect, he had become an unofficial authority. When the new Control Data was formed in 1957, most expected him to be part of it, and Cray was eager to join, but Norris stalled because he was worried about offending a potential customer. He knew that Seymour had been working on a special Navy project at the time and wanted to keep relations amicable. Seymour grew impatient. In September, he called Norris, saying, "I'm ready to come to work." He had invested $5,000 in stock in the new company, an amount he called "most of his assets," and he had ideas about using transistors to build computers. Norris asked if the Navy knew about his interest in joining the new company.

"Hell, no," said Seymour in his amiable voice, "I don't have to ask the Navy what I can do." Norris spoke of the problems it might create. Seymour listened and then in a stark showdown said, "Well, do you want me or not?"

"Of course," said Norris.

"Then I'll be at work tomorrow morning."

The Navy didn't give Norris the problems he thought it would, Norris said later. "They understood Seymour as well as we did."

Sometime in the following fiscal year, Seymour Cray was added to the company's board of directors, a level of authority quite unusual for a young engineer who was supervising only a small staff and reporting to vice president and board member Frank Mullaney. But everyone knew Seymour was a very special talent and should be treated accordingly. He had an unusual combination of mathematical and engineering expertise that allowed him to increase machine speeds with novel arrangements of electronic circuitry, while also developing methods to withdraw the heat generated by these more densely packed components. In short, he was a gifted architect. Or a composer. Speaking of his abilities, he once sounded like he might have the capabilities of Ludwig van Beethoven: "It is an art," he said, "because there is no logical way of proceeding with it. You have to put it all together in your head and hear it." Beethoven, the musical genius, worked with the same notes as everyone else; it was the arrangement of those notes that was spectacular. Similarly, the components Seymour worked with were available to everyone, but his arrangements were ingenious. And he had the boldness and quiet confidence of a genius. In today's parlance, he was a game-changer.

The purchase of Cedar Engineering had required a cash outlay for the young Control Data Corporation. At about the same time, Sperry Rand initiated a lawsuit against the company seeking an injunction against trade secrets, which required a costly legal response. Norris felt the need to conserve company cash, and in April 1958 key personnel were given a 50 percent cut in pay until equipment could be shipped and payment received. Because they had stock options, these employees had a special interest in the company's survival, and surely felt that if the short-term cash needs could be overcome, they might reap rewards later. They knew they were working on a technical development of great significance, even though the company's financial self-sufficiency was still in question. Norris and the founder/general manager of Cedar Engineering, E.J. Manning, cut their salaries to nothing and received notes in return. The information was kept quiet for competitive reasons.

In October, 1958, the company was able to sell $350,000 of 6 percent preferred stock to Allstate Insurance Co. It was a huge vote of confidence from an important Chicago financial institution to a little-known company just over a year old and not yet profitable. To some venturesome investors, it was now apparent that the ingredients of success for little Control Data were coming together. Allstate, because of its willingness to assume the early risk, also received a warrant to purchase 80,000 shares of common stock at $4.25 per share. The stock at the time was trading between $6 and $7 a share. But Allstate awaited further developments before capitalizing on its warrant, and that "kicker," over time, added enormously to its investment return. More than that, the structure of the arrangement may have become a model for venture financing by others soon to follow in the Twin Cities local market.

Within a year and a half, Seymour and his team of engineers built the prototype of a computing marvel that became known as the "sixteen oh four" (1604). It was "less costly than comparable machines, yet two to four times more powerful," Norris said. A limited executive operating system accompanied it, as sophisticated customers would be writing their own programs. Seymour, however, wrote the Fortran compiler used in scientific applications. The Navy took delivery of the first machine. Having received additional orders, the company could report to its shareholders in the annual report for the fiscal year ended June 30, 1959, that there was a shipment backlog of $5.7 million.

In February 1959, the two major Twin Cities brokerage firms published statistical reports on the company, but made no investment recommendation. On October 13, 1959, the *Minneapolis Star* commented on Control Data's "phenomenal growth." The stock surged to $17 that summer and $31 by calendar year end.

The next fiscal year the company had revenue of $9.7 million and net income of $552,000, or 55 cents per share. The company had found it necessary to raise additional capital as equipment was now being leased to customers. Norris ended his annual message to shareholders: "We face the future with optimism" and added that he looked for "continued profitable growth." Backlog stood at $11.9 million—more than double the level of a year earlier. Investors in Minnesota and other parts of the United States took notice.

On Sunday, May 29, 1960, Twin Cities media personality Cedric Adams wrote about Control Data in his *Minneapolis Tribune* Sunday news column. "There is a firm," he wrote, ".... one of Minneapolis' major electronic firms, [that] delivered its first machine last January and now has a total of four in operation, and is producing at the rate of one a month. The product carries a price tag of $1,000,000. I was amazed to find this production going on in a plant [501 Park Ave.] within half a block of our own." He went on to describe the many thousands of components inside the "room-size machine," and asked his escort, William Norris, "what the heck goes on inside that contraption?"

"Really, it's a pretty stupid sort of gadget," Norris replied. "All it can do is add." He must have paused, then, to let his listener feel superior in comparison. "But it must be remembered that all arithmetic processes can be reduced to addition. It can make a human being look pretty stupid." Cedric Adams went on to write that "the machine's decisions are thousands of times faster than those of a human being." He neglected to mention the stock or its price.

By July 1960, Control Data stock was trading at $46 a share. A few months later, the *Minnesota Business Development Newsletter* reported, "the stock has skyrocketed and the company has become a favorite topic of comment in the electronic trade from coast to coast."

During the next fiscal year, which ended June 30, 1961, continued demand for the company's products created the need for still more financing, so the company arranged a $10 million line of credit to support customer leases, and also sold stock to raise additional equity. By fiscal year end, the number of shares outstanding had risen to 1.1 million and the number of shareholders to 8,000, evidence of a truly active market. The company reported revenue of $19.8 million and net earnings of $842,000 or 73 cents per share.

Net earnings, however, were not a true measure of the underlying financial momentum then underway, as shipments placed on lease meant revenues would come in month by month, often taking forty months to generate revenue equivalent to a sale, while the expense of that equipment was depreciated under a "double declining balance method" over a four-year life, meaning 50 percent of its cost was written off in the first twelve months of the lease. This revenue lag, along with the aggressive depreciation expense, minimized the company's earnings in the early years of a lease. But depreciation is an accounting entry, not a cash transaction—the cash having been spent when the equipment was built—so in order to give investors a different understanding of the economics of the business, the company reported cash flow of $2.14 a share, almost triple the reported earnings of that year, and compared it to 80 cents a share in cash flow the previous year. This was a simple measure of incoming cash, principally earnings plus depreciation, but the additional information helped investors understand the company's underlying earning power. In summary, it could be said that Control Data was not only shipping many machines but was reporting earnings very conservatively (and legitimately, according to prevailing accounting rules). These were descriptors an average investor could understand, and may have given way to the thought, *Hmmm, understated. How much better must they really be*? The stock on May 29 of that year traded at $101.50, an achievement that merited special attention in the *Minneapolis Star*. Norris himself, having invested $75,000, was worth over $7 million, plus the value of whatever options he held, plus his wife's holdings. And the company was only four years old. At trading desks around town, the term "big datty" evolved. If "big datty" was strong on the opening, traders felt the market was going up; if not, it foretold weakness. Control Data had become an alpha stock.

Many local investors had bought the stock along the way, and didn't make a hundred times their money, but were still very happy. Norbert Berg, who was hired by the company in 1959, says he owed his father $1,700 on a car loan, which he repaid with 100 shares of stock when it was trading at $17 a share. His father later sold at $100 a share. "That was a very expensive 1957 Plymouth," said Berg. Home from military service, Berg's brother wondered where he might put some extra money, so Berg suggested he buy Control Data stock. His brother invested $450. Six years later he sold the stock for $12,000, almost the full purchase price of a house he bought in Wausau, Wisconsin.

Up in Hibbing, Minnesota, a dentist named Rudy Perpich got a call from a businesswoman in nearby Chisholm who had a toothache. It was Saturday, so he told her to come right over as he wanted to leave work and spend time at his lake home. When she arrived, he heard voices in the hall. A local stockbroker had stopped his patient and was trying to sell her some low-priced stock in a newly formed company.

Perpich became impatient and asked, "What's it going to be? Stock or your toothache?"

"Both," said the stockbroker.

"OK," said the lady, "buy the stock. I have to get my tooth fixed."

Years later, Governor Perpich heard from his former patient again. He learned that her toothache and the resulting investment in Control Data had made her more money than she earned in all her years in business.

Down in Rochester, Minnesota, George Waters referred to Control Data's success in explaining how he got involved with a company called FloTronics. "[Control Data] made an initial offering of a dollar a share, and everybody in town bought that stock. The next thing we knew, the stock was five, ten, eleven, twelve dollars a share. It made a lot of quick millionaires." Similar stories were being told all over the Midwest.

At $100 a share, the stock was trading at 137 times fiscal 1961 earnings, an extraordinarily rich valuation that would be vulnerable to a steep plunge should any operating difficulty occur. Seen in a different light, it also sold at 47 times cash flow, making the valuation seem more reasonable. Investors, knowing the business was doubling every year, assumed that the valuation on forward earnings was about half current levels, and half again in the year after that, quickly approaching more normal levels. They had heard that the company's computer, designed by Seymour Cray, was unsurpassed in operating performance, and that not only the government, but also universities and other scientific organizations, both domestic and international, were interested in placing orders. Such an expanding opportunity led investors to believe that revenues, earnings, and cash flow could continue to double annually for many years into the future.

Even the smart money out East was taking notice. Dick Jennison, a director of institutional research at the small New York firm of Auerbach, Pollak and Richardson, published an article in the *Commercial and Financial Chronicle* in July of that summer, recommending Control Data Corp. as "the security I like best." It was, he said, "the only small firm to penetrate the ranks of IBM, Sperry Rand, RCA, and Minneapolis Honeywell with a large scale digital computer...Under development is a giant computer many times the speed of [the current one] which may be priced in the $8-10 million vicinity... Studies indicate a sizeable market for its products in Western Europe... Subsidiary Cedar Engineering is developing peripheral equipment for outside sale." Whew! Growth in every direction. Jennison concluded that the company's "unusual qualities merit consideration by institutional investors as an attractive long-range growth situation." He did not mention the price; in other words, he was saying, buy it and don't worry about the very steep valuation even though you are an institution, careful in your approach and bearing responsibilities to others. This was a significant endorsement from New York. The chronicle carrying this recommendation had national exposure in financial circles.

Norris was invited to address the New York Society of Security Analysts, the high council of stock appraisers in one of the most important financial centers in the world. The former Nebraska farmer had become a figure of national reputation and he was prepared for this role, since he had already been to the New York area earlier in his career to explain the Univac Division to General Douglas MacArthur. On both occasions, he knew far more about his subject than his listeners.

Back in the Twin Cities, investors were able to say that Control Data was indeed the new ERA—even better—but with a difference. It was available to the general public, not just a few, and that had an enormous effect on the Twin Cities over-the-counter stock market. It created an enthusiasm that carried into the many stock offerings that followed, most priced at $1 a share just like Control Data—a price anyone could afford.

Norris ended his message to shareholders in 1961 by saying, "We are looking forward with confidence to future years of profitable growth," and numbers backed him up. Backlog stood at $24 million, once again approximately double a year earlier. The company began construction of a new headquarters building, three stories tall with two adjacent modules, at 8100 34th

Avenue South in the suburb of Bloomington, a mile or two from the metropolitan airport. Across the road, a dairy farmer was still milking cows. The lawsuit initiated by Sperry Rand Corporation four years earlier was settled out of court in January 1962. The future looked bright.

"I suppose we'll get big some day," Norris told the *Minneapolis Tribune*. Whether it was the understatement of an ambitious man, or the plaintive comment of an executive who knew the hazards of bigness, nobody could be sure. But to investors it sounded as though the opportunities in the computer industry were an unstoppable force pulling the company into a bright and prosperous future. Their minds began to levitate, imagining new riches. *If one company could do this, what about others?* Owners of businesses apparently imagined the same, and yearned to take their companies public and cash in on the riches. Those of an entrepreneurial spirit gathered a few key employees and created new companies often having to do with electronics in order to attract investor attention. Brokerage firms were now eager to do an underwriting. Soon new stock issues flooded the Twin Cities market, many at $1.15 a share (15 cents for the underwriter), and the public seemed eager to buy them all. They wanted the *next* Control Data.

2

Local Market, 1957–1959

New York, of course, was the center of finance in the United States in the 1950s, and its Wall Street district in lower Manhattan had come to take on an authoratative voice of its own. Just as Washington spoke for the central government and Hollywood spoke for the movie industry, Wall Street spoke for investment America. And because Wall Street—that is, the New York Stock Exchange and its less famous cousin, the American Stock Exchange—listed so many companies, the Dow Jones index was created to make it easier to follow the market as a whole. The index was quoted in points, not dollars, and there were separate indices for industrial companies, transportation companies, and utilities. There were thirty companies in the industrial index, representing perhaps a thousand others. Thus, a news report at the time might say: *Wall Street expected improving corporate profits today as the Dow Jones Industrials advanced 13 points.* And it was a summary of the nation's financial activity for the day. But such news was not widespread in the 1950s, because the general public was about as interested in the stock market as they were in opera, horseracing, or playing bridge.

Halfway across the country, however, out where agricultural markets were king, the Minneapolis/St. Paul stock market was coming alive. The quick success of Control Data caused such excitment—and investment speculation—that it might be said that this regional market had developed a voice if its own.

The function of the stock exchange, regional or national, is to maintain an orderly market. This is done by buying stock when there are no other buyers and selling stock (short) when there are no shares immediately available. In other words, the exchange fills the gaps in time and sometimes price between buyers and sellers. Stock movements in the amount of dollars were called "points," but share prices were quoted in increments of an eighth of a dollar—a practice that probably came from an earlier time when coins were exchanged in "bits," that is, eighths of a dollar (i.e. two bits for a quarter). Establishing such increments had its purposes. It eliminated nuisance movements of a few pennies, and it created a gap (spread) that allowed the market-maker who bought the stock to try to resell at a markup that left room for some profit. The market-maker had the risk of ownership and deserved to make some money simply for committing his capital. If only sellers showed up, the price moved down, and the market-maker lost money on the stock he was holding—his inventory. In performing this intermediary function, he was not much different from a small-time merchant who bought goods at wholesale, marked them up, and sold them at retail with the difference covering his cost of doing business, and if all went well, giving him a profit. Like any merchant, if the broker's goods went unsold, he had to mark them down and take a loss.

The New York Stock Exchange had locations (posts) on the floor where members set up shop, as it were, to buy and sell the specific stocks they handled. Without a market-maker such as this, an investor wishing to sell would have to find his own buyer, a difficult job for almost anyone. Thus, the existence of a buyer willing to accept the seller's shares gave the central exchange a very useful function, and because it provided centralized trading for each stock, there was usually a steady flow of activity and more stable pricing. Many New York brokerage firms had branch offices throughout the country, connected by telegraph, which brought in orders and added to the volume of activity at the various trading posts. Because of these networks, the New York firms were called "wire houses." In the 1950s, downtown Minneapolis and St. Paul both had branch offices of big-name firms such as Merrill Lynch Pierce Fenner & Bean; Paine Weber Jackson & Curtis; Smith Barney & Co.; and others.

But not all stocks qualified for listing on the New York Stock Exchange. Some had only regional interest or traded infrequently. Some companies didn't want to pay the fees required by the New York Stock Exchange. Nevertheless, shareholders in such companies might have a need to sell or a desire to buy. To meet this need, local brokerage firms opened offices to perform the function otherwise performed by New York stock exchange members. Such localized trading was called "over the counter."

These shops also quoted prices in eighths of a dollar, but sometimes activity was so unpredictable and so slow, or the dollar amounts so low, that the "bid" and "ask" quotations were as much as 30 percent apart. In 1957, for instance, tables in the *Minneapolis Tribune* quoted Northwest Public Service at 13.1 bid, and 17.5 ask. If someone had to sell, he might be glad to get 13 dollars 12½ cents per share because there probably were no other buyers. He could, if he wished, try to negotiate for a price between the stated quotes. A posted quote was generally good for a hundred shares. A Twin Cities stock such as Northwestern National Life was quoted at 103 dollars bid and none offered. In this case, brokerage firms were very reluctant to sell stock they didn't own for fear it would take months to replace, and then who knows at what price, so they quoted no offering price. Three weeks later, the same stock was quoted at 106 bid and none offered, the higher bid obviously indicating that someone was looking for stock. Two weeks later, the quote dropped to 88 bid and 98 offered. Clearly, stock was now available to be bought. And two weeks after that, the bid dropped to 84 and the spread narrowed to 6 dollars rather than 10. Stock could now be bought for 90, sixteen dollars less than the bid approximately one month earlier. A brokerage firm that bought stock at the recent high of 106 bid, and still held it four weeks later, would now be trying to sell it at 90, a loss of 16 a share. Such price changes indicate the risk of making a market in an illiquid stock.

In Minneapolis and St. Paul, the leading local brokerage firms at the time were Piper Jaffray & Hopwood, and J. M. Dain & Co., both of whom had wire connections to the New York Stock Exchange. Among the smaller companies in Minneapolis were Craig-Hallum Inc., M. H. Bishop & Co, Engler & Budd, John G. Kinnard Co., and Woodard Elwood Co., all trading primarily in local stocks or selling mutual funds. There were also one- and two-man shops such as C. D. Mahoney & Co., R. J. Steichen & Co., and J. P. Arms. Across the river, Harold Wood & Co., Irving J. Rice Co., Caldwell Phillips, Inc., and Sampair & Egan Co. were performing a similar service in St. Paul. Kalman & Co. had an office in both cities. The average stockbroker at the time was fifty-five years old, and many clients were probably older.

Local brokers offered a service, but it was not used by much of the public in the 1950s. It appealed to that element of the population interested in investing, and those few who had stock to sell, possibly inherited. Nevertheless, quality companies did list stocks in the Twin Cities local market. Below is a list of some significant names at the time and the prices, bid and ask, in the first week of January 1957, as reported in the *Minneapolis Tribune*.

	Bid	**Ask**
	(in dollars, fractions in eighths)	
American Hoist and Derrick	19.6	21.6
Brooks Scanlon	47	none
Donaldson Co.	11.5	13.1
Econ Lab 1	15	17
First Bank Stock	33.7	35.7
Green Giant B	21.2	24.2
Investors Div. Svs. A	64.4	68.4
Kahler	31	none
McQuay	7.5	8.5
Minneapolis Gas, com	26.4	28.4
No. Central Airlines	9	10
NW National Life	103	none
Otter Tail, com	26.7	28.7
Red Owl, com	29.2	31.2
St. Paul Fire	44.6	47.6
St. Paul Stockyards	19.6	none
SuperValu, com	33.6	36.6
Toro Mfg	21.4	23.4
Weyerhaeuser	34.4	37.4

There were 35 other stocks for a total listing of 54 companies. All but three were quoted above five dollars a share (bid), suggesting they had some substance behind them. It is worth noting that no stock carried a quote of less than a dollar a share, something that would not be true a few years later.

Minneapolis Gas, the local utility, was considered a reliable investment. It earned a better return than most other alternatives, and because there was a market, it could be sold when the owner needed to raise cash. Otter Tail, another Minnesota utility, was probably given the same consideration. First Bank Stock, Investors Diversified Services, Northwestern National Life, and St. Paul Fire were companies of size and durability even then. Their names evolved over time, First

Bank becoming U.S. Bancorp; Investors Diversified Services becoming American Express and then Ameriprise Financial; Northwestern National Life becoming Reliastar and later ING; and St. Paul Fire and Marine renaming itself St. Paul Companies and then Travelers Co. All are still major presences in the Twin Cities. Three other companies worth noting, because they kept their identities and continue to prosper even today, are Donaldson Co., Economic Labs, and Toro Manufacturing. The two large growth companies in the Twin Cities at the time, Minneapolis Honeywell and Minnesota Mining and Manufacturing Co., traded on the New York Stock Exchange. So did General Mills.

The average citizen was not overly concerned about stocks. He invested in savings accounts, postal savings, church bonds, and, to minimize taxes, municipal bonds. An income of $16,000 fell in the 47 percent federal income tax bracket. No doubt people heard about the money made in the stocks of Minnesota Mining and Manufacturing or Minneapolis Honeywell in the post-war years, but that was someone else's good luck. Most adults had lived through the Great Depression and it was still heavy in their memories, making them wary of speculation. If an ordinary American owned stock, it was probably in the company of his employment—an organization he felt he could trust. Some held shares in mutual funds, a "safer" alternative to individual stocks that was growing in popularity at the time.

Every morning a consolidated quote sheet was distributed to brokerage houses, banks, and other concerned parties in the Minneapolis and St. Paul downtown markets, giving the previous night's closing quotes on local stocks. This was one form of price communication. The other was the telephone, as traders had direct lines to each other allowing them to easily check competing quotes throughout the day. Those members of the public concerned about prices could call by telephone during trading hours, or go to a brokerage office over the lunch hour and check on quotes. Older men would often be seen sitting in a brokerage office amid a cloud of cigar smoke discussing the merits of one company or another. Those following a stock casually could check it daily in the newspaper. Almost all of the activity took place in the central cities, as the suburbs were still in the early stages of development.

A smart, attentive broker kept alert for relevant information. He was an immediate observer of price changes and picked up on new facts and rumors. If a stock was acting like it made sense for an investor to buy, the broker might pick up the phone, call a regular customer, and suggest action. The broker would sometimes know, for instance, whether a new quantity of stock was available to fill an order, or alternatively, whether an incoming quantity was so large it might indicate weaker prices ahead. He surveyed the financial terrain looking for capital opportunity. Almost all brokers were men. Women served as receptionists or "back office" bookkeepers. At the wire houses, women were hired to read the ticker tape and write the latest prices on an elevated blackboard covering one full wall for everyone in the office to see.

In 1957 the markets were quiet, and in January 1958 most of the stocks listed above were down only fractionally, probably in concert with the industrial slowdown then underway all across the country. The Dow Jones Industrial index was also weak during that period and continued to stay that way until the following summer. Investors Div. Svs. A, the same company that later built the signature IDS building in the center of downtown Minneapolis, stands out as the lone increase, up approximately 10 percent. The listing below shows the new (bid) price as of the first week of January 1958, and the dollar change from a year earlier.

	Bid	
American Hoist and Derrick	16.4	down 3.2
Brooks Scanlon	46	down 1
Donaldson Co.	11.4	down a fraction
Econ Lab	13	down 2
First Bank Stock	28.6	down 5.1
Green Giant B	16.4	down 4.6
Investors Div. Svs. A	72	up 7.6
Kahler	31	unchanged
McQuay	6.7	down a fraction
Minneapolis Gas, com	26	down a fraction
No. Central Airlines	1.6	down 7.2 (reflecting a split)
NW National Life	70	down 33
Otter Tail, com	25.6	down 1.1
Red Owl, com	29.2	unchanged
St. Paul Fire	44.2	down a fraction
St. Paul Stockyards	18.7	down a fraction
SuperValu, com	32.2	down 1.4
Toro Mfg	18.2	down 3.2
Weyerhaeuser	31.4	down 3

The latest listing had two additions not appearing the year before. They were:

Pacific Gamble Robinson	10.6
Apache Oil	6.1

These quotes gave every appearance of an orderly market consisting of quality stocks. But change was approaching, and it came from outside the normal underwriting activities of local brokerage firms. Officers in a new company, Control Data Corporation, had sold stock through their personal connections, and the company's quick success would soon re-characterize the entire local market. It would create a new set of expectations. Control Data would become the new standard of measurement and bring a different kind of investor into thc market.

In the mid-1950s, the residents of the Twin Cities were enjoying a welcome period of peace and prosperity. The Korean War had been settled shortly after President Eisenhower's election in 1952, and throughout his two terms the former general spoke in calm, reassuring tones that gave many a feeling of comfort and safety. He let the economy flourish or stumble on its own with minimal government interference. Though he did send federal troops into Little Rock, Arkansas, to accompany black children into an all-white high school in an early confrontation over desegregation, most of white America considered civil rights somebody else's problem. They were busy buying houses in the tracts of first-ring suburbs and filling them with labor-saving appliances like vacuum cleaners, washing machines, and dishwashers. Wives stayed home and looked after their children while husbands left for work every morning seeking their share of the new prosperity now seemingly accessible to all. At home in the evening, couples enjoyed the new medium of television.

But because they remembered an earlier period of war and depression, all adults knew that prosperity might be tenuous, so they worked and saved and spent prudently. Twin City Federal (now TCF Bank), a savings and loan company headquartered in downtown Minneapolis, grew to huge size by urging people to save, and later had an advertising slogan suggesting that the public "tuck-a-buck-a-day-away." The rhythmic line brought in many millions. And the money saved and deposited in their offices was, in turn, lent out for mortgages against the many homes being built and bought around the Twin Cities.

There were no freeways. University Avenue was the main corridor between the downtowns of Minneapolis and St. Paul. Overall, traffic flowed in an orderly way, unless the Minnesota Gophers were playing football on the campus at Memorial Stadium, which held 60,000 fans, or the Minnesota State Fair was underway in St. Paul.

By 1954, streetcars had been retired in favor of buses. But in Minneapolis, a large train depot still stood at lower Hennepin Avenue, along the Mississippi River, from which the Great Northern Railroad sent passengers west on their way to the Pacific. A few blocks down Washington Avenue, the Milwaukee Road carried passengers to and from Milwaukee, Chicago, and connections east. In between those two hubs was the Gateway District, an intense concentration of bars serving cheap drinks, rooming houses offering an overnight stay for fifty cents, and an occasional strip joint. Missions in storefronts had speakers preaching the Gospel while others served bread and soup. The brick buildings lining Washington Avenue retained the commercial look of an earlier period, but the sidewalk now teemed with men living alone, many of them migrants from elsewhere, near and far. During the day they sat on benches or ambled around looking for handouts. A person with money was not safe in the area at night. The cops gave warning: *drunks might roll you and take your money for liquor.* Periodically an influx of work gangs from the railroad, called "gandy dancers," arrived in town, eager to spend their paychecks on liquor and the city's sensual allure. During the winter they collected unemployment. Fist fights might break out when arguments got too intense, and sometimes there were knives or blackjacks, but usually no guns. It was a rough and crude area soon to be demolished in the Gateway Redevelopment effort.

Beyond this zone of urban decay lay a modern urban core surrounded by many miles of lakes, parks, and comfortable homes. The Foshay Tower, an obelisk of thirty-two stories, rose above the central city, double or triple the height of surrounding buildings. It seemed a sentinel, as if to say, *this is the start of something big,* but it stood alone. It had been built in the final days of a prosperous decade and opened in 1929, just as the Depression was about to begin. Now it looked somewhat forlorn, too tall for its neighbors. Nearby, Northwestern National Bank had a large round sphere called a weather ball above its building, lower than the Foshay peak but still visible from fifteen miles away. It changed color to indicate the coming weather—red for warm, white for cool—and flashed on and off to indicate precipitation. This was a more useful symbol than the obelisk, as it showed that the city underneath was alive with workers who wanted to know how warm to dress for work or an evening out. Dayton's department store (now Macy's) faced the corner of Seventh and Nicollet, and across the intersection stood Donaldson's, its principal competitor. J.C. Penney, Powers, and other stores also sold clothes and general merchandise. Both Kresges (forerunner to Kmart) and Woolworth's, then known as "five and dime stores," sold variety items. A shopper had choices, all within a few blocks of each other. The IDS tower was still fifteen years away.

Charlie's Café Exceptionale, a two-story English Tudor building on Fourth Avenue and Seventh Street, offered dining elegance on linen table-cloths, a variety of steaks, fresh bread, man-sized drinks "expertly mixed," flaming desserts, and an ambiance agreeable to lingering discussion and an after-dinner smoke. It had various dining rooms and could feed up to 350 patrons. At the other end of downtown, at the intersection of Hennepin and Washington avenues, the Waikiki Room inside the Nicollet Hotel served meals in an exotic Hawaiian setting with Polynesian carvings. This was a somewhat unconvincing indoor simulation of the real thing, but it was a nice place to eat and to enjoy Polynesian rum drinks at a time when most Minnesotans could not easily afford to board an airplane and fly out over the Pacific. Both buildings have since been demolished. But Murray's Restaurant and Cocktail Lounge, popular then as a steak house and a gathering place in the center of downtown, remains today as one of the few surviving local businesses from that era.

The University of Minnesota had not yet extended its campus across the Mississippi, and the river's West Bank was crowded with the dilapidated housing of Bohemian and Scandinavian communities and stores and taverns at nearby Seven Corners.

The principal industry in Minneapolis at the time was flour milling. General Mills, Pillsbury, International Milling, Archer Daniels Midland, and Peavey all were dependent on the railroads that brought each fall's harvest into the city from the vast grain fields of western Minnesota, the Dakotas, and Montana. Cargill engaged in grain trading and stored grain in rural elevators. Prices for the grain were set at the Minneapolis Grain Exchange, then one of the country's largest commodity trading centers. Financing all this were two bank holding companies, First Bank Stock Corporation (now U.S. Bank) and Northwestern Bancorporation, also called Banco (later Norwest, now Wells Fargo). Each had numerous subsidiary banks in the smaller towns throughout the agricultural area to the west.

Minneapolis Honeywell Regulator Co., headquartered on the city's south side, grew large during World War II on military production, and after the war continued to manufacture flight controls, switches and gyroscopes, thermostats, and other electronic innovations. It was the state's largest private employer in the 1950s with 14,000 people in twenty-one locations. Annual sales volume was approximately $300 million in 1957.

The activities of the city, in short, were oriented west toward a newer America—a vista of agriculture, finance, emerging technology, and the prosperity that accompanied those pursuits.

Downriver, around a fish hook bend in the Mississippi, stood the older city of St. Paul, developed in its early days from river traffic and later, railroads and lumber. Now it was identified by a triangle of three prominent buildings about a mile apart. The first stood near the river in the center of downtown. It went up thirty-two stories, even higher counting the sign above the top floor with an elongated number 1, followed by "st," indicating the First National Bank of St. Paul and its sister company, the First Trust Company (now both part of U.S. Bank). It seemed to say, *this is the center of town.* The sign was bright red and could be seen twenty miles away.

Uphill to the north stood the Minnesota State Capitol, a white marble building of massive shoulders and a sculpture of four golden horses about to dash off its roof into miraculous flight over the city. A dome rising in the center gave the building a regal appearance.

And on a hill to the northwest, another dome reached to the heavens, its elevated cross indicating a cathedral.

Reflecting an early immigrant desire for education and upward mobility, there were six private colleges in or near the city, each associated with a European Christian religion: St. Thomas (Catholic men), St. Catherine (Catholic women), Macalester (Presbyterian), Hamline (Methodist), Concordia (Lutheran), and Bethel (Baptist). Northwestern Bible College at the time was located in Minneapolis.

On the east side of town, Minnesota Mining and Manufacturing Co. (3M) had become a surprising corporate success with products such as sandpaper, copying technology, and Scotch Tape. The company stressed innovation. Sales volume reached $330 million in 1956. Its magnetic tape operation, located 65 miles away in Hutchinson, Minnesota, was noted for automation, as it had only 150 employees. The company was progressive, very much focused on the future.

Straddling downtown St. Paul were two large brick breweries, Hamm's and Schmidt's. They stood for an older, blue-collar history and a less inventive economy that was still very much alive. Competing on a national level, Hamm's television commercials featured the musical line "from the land of sky blue waters," an allusion to the state's pristine northern lakes. An angel voice, pure as cream, sang the memorable ditty while tom-toms beat in soft accompaniment, almost an Indian love song. Alas, both breweries ultimately succumbed to the squeeze of relentless advertising from larger national brands.

Many three-story homes with grand porches across the front and stables behind (now converted for cars) faced Summit Avenue and other streets, especially in the Cathedral Hill area. They evoked a time of comfortable elegance and gracious living perhaps best captured in works by the novelist F. Scott Fitzgerald, who once lived there.

Thus, both Minneapolis and St. Paul prospered on the strength of established institutions and businesses. But there was also evidence of a coming surge in prosperity based on new ideas and new technologies. An individual owning Minnesota Mining and Manufacturing (3M) stock at the end of World War II, now had sixteen times as many shares because of splits, and had a roughly commensurate increase in value counting appreciation and dividends. Minneapolis Honeywell Regulator stock had gone up tenfold in the postwar years (the name was officially shortened in April 1964). Investors Diversified Services (IDS), the Minneapolis investment management company selling mutual funds and savings certificates, saw earnings swell from the rising market of those years and from the upward creep of interest rates. Higher stock values meant higher fees, and rising interest rates meant a higher spread over interest paid. Its Class A stock went from 14 bid, 16 offer in the summer of 1950 (the first year of quotes listed in the *Minneapolis Tribune*), split 5 for 1 four years later, and by the summer of 1957 traded over 100. That amounted to a thirty-fold increase from the offering price of seven years earlier. During much of the first year, however, there was no offering quote, so it was difficult to find a willing seller. Those lucky enough to do so rode an escalator to easy riches.

3

Hot Stocks

It began slowly. Control Data had brought new excitement to the market, and it seemed like an easy example to duplicate. A few new issues came public and went up in price. Someone who made money on an issue would tell his co-workers, his family, his friends and neighbors; he'd tell them whom to call for the next new issue. "Just buy it, and sell later, if you want. I hear it's hot." A man who's made easy money has a hard time keeping quiet.

Brokers kept lists. "I'll try to get you some, but I have to take care of my regular customers. Do you know any other people who might be interested? There's another issue coming next week. I think I might be able to get you some."

"Should I buy that too?"

"Yes, I expect it to be hot."

HOT. Soon everything was hot. That meant it was going to go up in the aftermarket. It was on allocation at the initial offering and there would be follow-on demand. The process was self-fulfilling: Because an offering was considered hot, it became hot. Everyone wanted a piece of it.

Preceeding the frenzy, there were five new offerings added to the Twin Cities over-the-counter (OTC) quotes listed in the *Minneapolis Tribune* in 1958. Nothing exceptional, it would seem, although some turned out to be substantial companies. Only one, Glass Products, appeared to be a cheap speculation.

Control Data	11.6 approximate year-end bid
Glass Products	1.7
NW Banco Pfd	116
Schjeldahl	8
Thermo King	16.5

In the first six months of 1959, two more companies were added. Again, nothing exceptional, although one of them, Patterson Dental, went on to become a large, successful company that continues to thrive even today. The founders of the other, Data Display, were engineers who left Univac to build large, computer-driven cathode ray displays, primarily for the government. *Hmmm, hadn't that happened once before? Better buy that,* some probably thought, *or at least watch it.*

Data Display	3.3 approximate mid-year bid
Patterson Dental	13.4

In the last six months of that year there was a small explosion of thirteen new listings. Nine of them ended the year with a quote of $5 or less, most likely up from their initial offering price in the $1 to $3 range.

Am. Electronics	1.4 approximate year-end bid
Apache Realty	3.1
Gen Elect Controls	.5
Gen Magnetics	1.1
Jostens	21.6
Layghton Paige	4.5
Midwest Tech	5.3
Mpls Assoc	3.4
Pik Quik	3.3
Research Inc	4.2
Telex	17.4
Waco Porter	6.7
Washington Mach & Tool	2.7

And in the first six months of 1960, thirteen more new issues appeared, seven of which were quoted at $5 or less. One of those listings, a company named Medtronic, made its introductory appearance at 2.4 bid, 3.4 offer, in the month of May.

By midyear, new issues named Baker Eng, Commercial Resins, FloTronics, Marquette, Space Structures, and Transistor Electronics all were trading at prices higher than Medtronic; perhaps investors thought they had better prospects.

And they kept coming. In the last half of the year, twenty-four more new companies were listed, seventeen of which ended the year priced at $5 or less. New organizations popped up like prairie gophers in the summer sun, one every ten days or so. Virtually anything could be brought public in 1960, including the following abbreviated names as listed in the *Minneapolis Tribune*:

Arrivals
Comm Chem,
Durox
Eagle Wash
Int. Prop
Mars Industries
Minn Elect
Minn Pharm
Servo Eng
U S Bowling

All traded under $2 at year end. The next Control Data, indeed!

In the first half of 1961, another forty-two new companies were listed, and in the last half of the year, another fifty-four. Amazingly, all ninety-six got financed, even though they arrived at the rate of two a week in the later period. No doubt brokers were feeding the frenzy, but the public participated willingly, sometimes standing in line to pay their bill. The new-issue market was ablaze.

Perhaps it received a psychological boost when the Russians launched Yuri Gagarin into spacc on April 12, to be followed a month later by the American Alan Shepard, as investors reasoned that business would surely benefit from new space spending. But for the most part, market exuberance appeared to be a spontaneous outpouring of capitalism, though Control Data's striking success certainly contributed to the widespread optimism. As might be expected, a measure of foolishness and impropriety accompanied the enthusiasm.

When introducing a new offering, some brokers learned to speak softly into the phone, making the customer listen more carefully. The muffled tones made the words sound like a special opportunity. *I've got one here I think you'll like. I don't want everyone to know about it. Can you hear me? I expect this one to be hot.* Sometimes they were right. Certainly they were right enough times to keep the customer coming back. And after selling one or two at a profit, the customer felt like he was playing with free money.

"How much should I take?"

"I'll try to get you a thousand. How about your brother? Is he interested?"

"Yes, and I think my neighbor too."

Les Bolstad, Jr. had a job rolling sod in 1960. He was nineteen years of age, making 95 cents an hour, and coming home tired and dirty every night. Then he got a job at Northern Ordnance paying $1.65 an hour for cutting the grass—a nice move, he thought. It was easy work, with coffee breaks, and he was told to stretch the work to make it last longer. Bob Roosen, a friend, was working in the brokerage industry. The story on Roosen was that his father had offered him a chance to either go to college or to buy Control Data stock. Maybe he chose both, since he and Bolstad were college fraternity brothers and he also became an enthusiastic promoter of local stocks. He helped form the Continental Securities brokerage firm. One day Roosen asked Bolstad how much money he had. Bolstad told him, and Roosen said, I'll buy you some stock, send you the confirms. Bolstad thinks he may have bought Control Data, General Electronic Controls, Telex, and maybe others, he's not sure. "Wasn't even sure at the time," he says. But he made money and was hooked. He decided to get a securities license and go to work for Continental Securities. He would receive a $250 a month draw against commissions—that's $3000 a year guaranteed, less than his previous pay, but it had upside possibilities. He was given an office, a chair, a cardboard box for a desk, and a phone with buttons for line one and line two—advanced technology at the time, making it possible to speak to one customer while placing another on hold. As the lowest ranking employee he sat at the trading desk when the company officers went to lunch. "Don't buy anything," he was told, "Don't sell anything, either," and then a final summation, "Don't lose any money."

The firm decided to underwrite Product Design and Engineering, a company making blow molded plastic products and stamped and machined metal parts and tools. The company had been doing business in Golden Valley since 1955, and sales were running about $400,000 a year. Before the offering, current liabilities exceeded current assets by $22,000, but the new offering would bring $225,000 in new cash, with stock priced at $1.15 a share (15 cents to the underwriter). Les was asked to work on the offering. Somehow, the deal got hot. He sold many thousands of shares. "You would get one guy at Univac to buy, and five more would call. Anyone could gamble $115 for 100 shares," he said. David Lundstrom, then working at Univac, saw it from the other side. "Many engineers had their broker's telephone number prominently posted above their desks. A typical engineer at Univac might own 100 shares of a half dozen small companies. Some engineers organized investment clubs that met monthly and purchased mostly local stocks." One engineer had a graph over his desk that showed the price of Control Data stock on one axis, while the other axis was labeled Number of

People I Have To Be Nice To, with diminishing numbers as the price went up.

"All I had to do was answer the phone," said Bolstad. "Made $5,000 in a few weeks." Then his firm hired more sales people. He says he never had another deal as good as that one. The stock, Product Design, bounced around the $1 range for seven years and then ran to over $8 a share in 1968 and even higher in later years. One could argue that investors weren't necessarily wrong in buying it, just seven years premature.

"It was capitalism at its lowest common denominator," said Bolstad. "Most people were honest." The firm bought a recorder so customers could call, listen to quotes at the end of the day, and pick their own selling point. "Cowboy capitalism," he called it. "There was no analysis." After work, brokers would get together at Busters on 6th Street in downtown Minneapolis and swap stories over booze and smoke.

Convention Centers, Inc. was another new issue. It was formed in 1960 under the original name "Minnesota Vikings, Inc." but the name was relinquished when local owners received a franchise from the National Football League. The company instead received a franchise to the National Bowling League. It owned seventeen acres adjacent to Metropolitan Stadium in the suburb of Bloomington and intended to construct a ninety-six-unit motel and an arena to be used for bowling matches, meetings, concerts, and other convention activities. The offering was completed in early 1961 at 650,000 shares for $1.15 a share. Max Winter, who would later become an owner of the Minnesota Vikings professional football team, was listed as executive vice president of the company. The stock ended the year at a quotation of .3 bid, and a year later it was gone. Although investors ended up with nothing, the cash raised had been used to build a convention arena called the Metropolitan Carlton Club, which served as an entertainment venue for a number of years thereafter. Some shareholders may not have felt too badly about the loss, however, as they "almost" owned a piece of the Minnesota Vikings. The National Bowling League, on the other hand, did not fare as well.

A college student during those years, Pat McNeil was working part-time at Minneapolis Honeywell. He remembers lunch breaks with other technical workers as they talked about the money to be made buying one stock or another. Usually they had just spoken to their brokers and had the latest prices. It was speculation more than investing, and the fever had spread into many Twin Cities factories. Sometimes those factories gave impetus to a group of workers eager to leave, form their own company, and capitalize on a rising stock price.

Out in Seattle, Washington, a brokerage firm named Blanchett, Hinton, Jones & Grant was selling stock in a company called Rocket Research Corp. It had been formed by four former employees of the Boeing Airplane Co. who had been engaged in rocketry and propulsion. They intended to do similar work in their new company. Proposals had been submitted to government agencies, but none had resulted in contracts, and no industrial products had been fully developed. Most likely aware of the ease of raising money in the Twin Cities, the brokerage firm teamed up with Craig-Hallum Kinnard Inc. (formed by a merger; the two names eventually split). An offering of 300,000 shares was undertaken at $2.25 per share. It was successful and the stock traded at 3.4 bid at year end.

On March 25, 1961, *BusinessWeek* published an article saying that across the country "small investors have been bitten with the dream that stocks they buy now for a few dollars will someday be worth millions." A new generation, too young to have been permanently scarred by the Great Depression, were now putting their trust—and dollars—into the stock market. The article gave special attention to the Minneapolis market, which had "caught fire from Control Data's success."

A few blocks from downtown, Eddie Fleitman helped get the *Minneapolis Tribune* out every day, working with pressmen and truck drivers. Sometimes he went around collecting coins in the newsstands. It required a strong back and diligence more than an education, but, as in most jobs, personality helped. Eddie had personality. "I hated school," he said, "and I had a family to support." The job with the newspaper enabled Eddie to get around the downtown area and keep up on what was happening. When the local stock market started to bubble, Eddie knew about it and spent time among the various brokerage houses clustered around 6th and 7th streets, near Marquette and 2nd Avenue, the city's financial district.

One day he bought 100 shares of a new issue at $1.15 per share. When the stock opened for trading, Eddie sold at $1.65 a share. It was "scared money," he says. But he was delighted at the easy fifty bucks. He followed the talk about other new issues and kept participating until one day a more expansive thought sprang into his mind, *what if I bought 200 shares?* Again it worked. He told his fellow workers about the quick riches and some of them began telling him they wanted to place orders. He told them whom to call to get the stock they wanted.

Increasing demand now meant that brokers could allocate stock among their customers. Because of Eddie's many referrals, brokers took pains to make sure that he always got his order filled on a new offering. Almost all new issues were then in demand as all were expected to open at a higher price when trading began.

One day Eddie's broker at Naftalin and Co. said, "Eddie, go to the bank, get some money and I'll give you 1,000 shares." Demand was now fueling demand, and capitalizing on it made sense. If a stock was going up, own more.

Eddie was not the only one aware of these recurring windfalls. He says there were sometimes lines of people out on the sidewalk waiting to buy or sell when trading opened. And who were the people driving up prices by buying in the aftermarket? Eddie wasn't sure. Perhaps they were those unable to get stock on the initial offering. Sometimes when a broker offered to give a buyer the shares he wanted at the initial price, which was considered a good deal, the broker expected the customer to buy more once trading began. This, of course, helped create the aftermarket. With his increasing investment activities Eddie soon knew most of the Minneapolis brokerage houses that specialized in low priced new issues. These included Naftalin and Co.; P. R. Peterson; Continental; York and Mavroulis; Corporate Securities; and Bratter and Co. in Minneapolis. Across the river in St. Paul were Brandtjen and Bayliss; and W. R. Pewters and Co. J. P. Penn and Co. had offices in both cities.

After learning about stocks during the wild excitement of the early 1960s, Eddie Fleitman went on to become a broker at Kinnard and Co. and then a sales manager at Craig-Hallum, Inc. "Eddie was very effective," said George Bonniwell, his boss at Craig-Hallum, "Everybody loved Eddie." And that may have counted more than anything else in the local over-the-counter market. "Even Carl Pohlad became a customer," Eddie said. Through his banking business and other enterprises, Pohlad eventually became one of the wealthiest men in Minnesota, but Eddie didn't claim credit for that. "All Carl asked was to be treated fairly," he said.

In spite of the broad choice of brokerage firms, many businesses seeking financing chose to underwrite their own stock issue, just as Control Data had done. But once the shares were in public hands, a brokerage house or two would generally agree to make a market in the stock. Who could predict the future? Perhaps an obscure new company would be a surprising success.

Many of the small brokerage firms had been newly formed as local stock activity entered ordinary conversation and the money-making possibilities drew the attention of enterprising individuals. Disdained by more experienced firms for their opportunism, they were called "bucket shops," the term suggesting a company with no more accounting sophistication than two buckets: one for "buy" tickets and one for "sell."

It was fairly easy to open a brokerage firm, Howard O'Connell remembers. He was a state securities examiner and later a brokerage executive. He remembers the price at $300 for a license, and a $5,000 minimum capital requirement. John Steichen remembers the same, although his firm, R. J. Steichen and Co., named after his father, had been in existence for a number of years by the time Control Data arrived. There was also office rent to be paid and the expense of desks and telephones, a total amount perhaps similar to opening a small grocery or a simple restaurant. Once a few brokers were hired and they brought in friends and relatives as customers, a firm could quickly generate enough revenue to cover expenses, especially if it could participate in new issues. Some new brokerage houses such as John Stephens and Co. and Rockler and Co., kept overhead low and earned their money primarily by trading on the spread—that is, buying on the bid and hoping to sell at a higher price for those stocks in demand, or alternatively, selling at a price and hoping to buy (or get flat in the position) at a lower price on those stocks showing weakness.

Customers found it expedient to open accounts at several brokerage houses in order to participate in their new offerings. Brokers differed in their areas of expertise and insights into local investment opportunities, so maintaining several accounts made good business sense. George Kline, a 6'4" former University of Minnesota star basketball player, was a young employee working then in the local securities industry. His coach at the university,

Ozzie Cowles, liked to invest in local stocks; he liked the action, George said. One day, Cowles was at the firm, talking about FloTronics, a stock he thought looked good. His broker agreed, and after some discussion, Cowles said, "Buy me 500 shares." After a few minutes, he was told his order had filled. "Oh, no!" said Ozzie, "I just bought my own stock." He'd forgotten he had a sell order at another house.

George Kline became sales manager at Naftalin and Co., even though, by his own words, he "knew very little." He would bring in company executives to talk to the brokers. One day he found himself on national TV, live on the *Today* show, which was normally broadcast from New York. They had brought a news anchor right into the office as part of a segment telling the nation about the "red-hot Minneapolis/St. Paul local OTC securities market."

Jim Fuller, a reporter for *Electronic News*, was a close observer of this feverish financial activity. He decided to start a magazine devoted to thoughtful reporting and the regular publication of significant information and news. His first issue of *The Upper Midwest Investor* appeared in April 1961. In the lead editorial he stated that "it is our belief that many persons now investing in local securities are doing so without adequate knowledge—not generally because they are careless, but because the necessary information is frequently unavailable." The opening issue already had letters to the editor. One said that the "local rumor mill" had been the chief source of information, and another said the most frequent guide to local investing had been the "hot tip." An article within the body of the magazine reported that Governor Elmer L. Andersen had appointed an advisory committee to investigate the recent "new issues" boom in the local securities market, and went on to say that some of the larger local brokerage firms had "forbidden their salesmen to make recommendations on most local securities." Though it highlighted these concerns, the magazine's underlying intent was to preserve the environment of funding that was helping build a new economy. There were feature articles on various local companies, outlining their competencies, weaknesses, and possible future prospects. *The Upper Midwest Investor* quickly developed a reputation for reliable, objective reporting.

Protecting the general public from reckless enthusiasm may have been the principal motivation of those expressing concern about the seemingly chaotic state of local investments, but competitive envy might also have been a factor. The older, prestigious brokerage firms had a carefully cultivated reputation of providing sound investment advice, and they were losing business to upstart competitors willing to underwrite almost any company with a possibility of success, however remote. Because so many new issues were brought public in the dollar price range, the term "bucket shop" took on another meaning for some: *selling stock at a buck.*

New issues sold directly to the public did not receive the third-party review for quality that an experienced investment firm could provide, nor did they receive the coaching and interim financial help they might need until they were ready to stand on their own. The public was expected to make its own judgments about a company's prospects.

All offerings were reviewed by the state securities commissioner, however. In an interview in 1961, Commissioner Arthur Hansen estimated that his office rejected about one out of four submissions. "What the public can be assured of is that the commission has not detected any evidence of fraud," he said, adding that "the new formations have added much to the prosperity of the state." He cited an electronics firm formed four years earlier that currently employed more than a thousand people.

To the older brokerage firms, undisciplined financial underwriting arrangements threatened the reputation of investing in general. It had moved out from under their professional oversight. Further, the sources of money for these new issues had spread beyond people of financial means to the public at large—to the factory worker, to the unsophisticated, to the impulsive speculator, to the money-grubber. Older authorities felt a responsibility to protect these beginners from their own foolishness and the harm that was likely to follow.

Arthur Hansen

In a subsequent issue of *The Upper Midwest Investor*, a reader wrote of suspected "market rigging," adding that local brokerage houses were "guilty of hogging" most of the new issues for themselves. "Hot issues are always oversubscribed, no matter how far in advance customers place their indications," the writer said. "This is because the brokers buy up all the stock for their personal accounts at the offering price; the stock opens at triple the original price; the brokers then unload their stock to the public at greatly inflated prices." Another reader wrote: "I have heard of many, many cases of manipulations and unfair dealing of one kind and another in our local securities market. Why isn't something done to stop these underhanded practices?" The magazine's editor, Jim Fuller, admits today that there were abuses going on at that time, and people knew about them, but they weren't willing to name names. It seemed investors were upset about the circumstances of their individual cases, but didn't want to threaten the overall environment of easy money.

Robert Smith, principal in the St. Paul firm of the same name, summarized the activity of the time in an article published in *ACE* magazine:

Whether you realize it or not you have recently witnessed the most fantastic period in local financial history. What has happened in our local securities market in the past two years, in many respects is far more unbelievable than was the 1929 stock market boom.

In '29 if a stock went up 100% in six months it was quite an event. Yet in our local market, this year many of our low-priced stocks went up 100 to 600% in six days to six weeks. And a new issue that only went up 50% was quite a disappointment.

He traced this "strange development" back to the formation of Control Data at $1 a share just four years earlier. Now, he said, nationally known securities analysts and financial writers complain that they are receiving many inquiries regarding these securities, but that in many cases the stocks have risen another 20 to 50 percent before they can answer. Describing the "exasperating speculation," he quoted a national source, Jack Dreyfus, president of The Dreyfus Fund.

Take a nice little company that has been making shoelaces for forty years and sells at a respectable 6 times earnings ratio. Change the name from Shoelaces Inc., to Electronics & Silican Furth-Burners. In today's market the words electronic and silicon are worth 15 times earnings. However, the real play in this stock comes from the word "furth-burners," which no one understands. A word that no one understands entitles you to double your entire score. Therefore, we have 6 times earnings for the shoelace business and 15 times earnings for electronic and silicon or a total of 21 times earnings. Multiply this by 2 for furth-burners and we now have a score of 42 times earnings for the new company. This is simple; anyone can do it.

At the macroeconomic level, however, something very significant was occurring because of this crazy behavior. Willis

(Bill) Drake, one of Control Data's founders, remarked in a guest editorial in *The Upper Midwest Investor* that various measures already indicated that the area "has grown to be one of the major technical industry centers in the nation," and that it was probably "the availability of risk capital, *more than any other single factor*, that made this possible" (italics added). He called on all involved to nurture this prosperity, but also to protect it with "meaningful and full disclosure and the availability of competent advice and counsel." He described how such a fertile environment came about.

> *There has evolved here, partly by design and partly through good fortune, a unique environment which has facilitated the growth of our technical industry to this point in the amazingly short period of three years. Included in the excellent blend of helpful factors are technical manpower, an excellent university, the interest and support of many business leaders and the press, ample risk capital, and the most fortunate presence of some ambitious, technically competent entrepreneurs. While each of these factors is vitally important, it's the combination that is winning.*

In April 1961, Continental Securities brought out a new issue that was typical of many at the time, intending to raise $300,000 at $1.15 a share. It was done on a "best efforts" basis; that is, there was no commitment by the underwriter to buy any unsold stock to make sure the offering would be successful; it would simply be sold to the public in the best effort possible.

The company was called Nucleonic Controls Corp., a Space Age name certainly meant to inspire respect. The prospectus, only ten pages long including financial statements, had narrative in large type, easy to scan. There was a balance sheet but no profit and loss statement as the company had no operating history. It did have a president of some credentials, however—Maclean R. Brown, an M.I.T. graduate in electrical engineering who had been "associated with General Electric" for twelve years. Following that, he was "a manufacturer's representative dealing with products in the public utility field." His new company had elaborate plans. Although there appeared to be only two full-time employees, it intended to set up seven operating divisions engaged in the following activities:

Electronic Rot Detector Division, which would manufacture an "apparatus" able to detect "internal rot in wooden objects such as power line poles, telephone poles, bridge trestles, live trees and the like. A working model of the apparatus has been developed."

Repair and Service Division, which will service and repair the rot detectors.

Pole Survey and Consultant Division, which will do pole surveys including a complete "history on each pole... to be recorded on data processing cards."

Data Processing Division, which will "compile the information received from actual pole surveys," allowing the customer to "replace poles under a planned schedule at a cost more reasonable than the replacement of poles under emergency conditions."

Manufacturing Division, which will—you guessed it—"assemble and inspect" the rot detector, and also "handle the manufacture, production and inspection of other products now being considered."

Research and Chemical Analysis Division, which will "analyze the most effective method of treating the poles," and also "develop new products."

Sales Organization, Advertising and Promotion Division, which will handle the advertising, promotion and sales of the "rot detector and the pole survey." In addition, it will be the exclusive sales agency for "pole top breakers and switches, instrument current transformers, and yard light luminaries... in the United

States." These additional products were made by General Magnetics, Inc., a small Minnesota corporation, not to be confused with Magnetic Controls.

Maclean Brown was to receive a salary of $25,000 a year, probably for overseeing the seven divisions. The other employee would be paid $12,000 a year.

At an investor presentation in a private room over lunch at McCarthy's Restaurant, Mr. Brown demonstrated the apparatus on a pole carried into the room for the occasion. Those who were at the demonstration were amused to recall that the apparatus ratcheted up the pole, and upon detecting interior hollow rot, let out a "ding–ding–ding" like a slot machine announcing a jackpot. Brokers, in trying to sell the offering, suggested to customers that they focus on telephone poles when next driving down a country road. It would be like bending the teeth on a comb and watching them wave back into position. There was a market for these things, right? And that didn't even consider the number of trees in the nation's forests rotting away undetected. Or foreign markets, like Canada.

The money was raised and the company stumbled along spending its cash until eventually Howard O'Connell, then the president of Continental Securities, arranged to have the remaining shell of the company become part of Analysts International, a contract programming business. Analysts International had two board members of earlier note: Frank Mullaney and Bill Drake, both founders of Control Data Corporation; Fred Lang, the president of Analysts International, had worked with both of these men at ERA. The business of detecting rot in telephone poles was abandoned, and the shareholders of Nucleonic Controls had another chance, this time in the business of computers, something everyone knew were being used by American business in ever-expanding applications. Shareholders were fortunate in having an underwriting firm that took responsibility after the initial effort proved futile. The stock hit 20 at the end of 1967, up from 1.15 in 1961. Many other new issues were not so fortunate.

And so it was with the local stock market of that period, a lottery pick. Control Data had risen to 100. Who knew what else was possible? But in order to win, you had to play.

As one of the founders of Control Data Corporation, who held the title of vice president, secretary, and treasurer, Arnold (Bud) Ryden could have cultivated that opportunity over the next ten years and ridden it to a mountain of riches, but Ryden was a "player" with an MBA from Harvard. He was impatient and ambitious. Most of the other principals involved in Control Data were electrical engineers who knew they were on the verge of a product opportunity. They were moved as much by the excitement of technical innovation as by money. Ryden, it was generally believed, wanted the company to expand and grow by merger and not be so dependent on government contracts. Bill Norris, president of the company, apparently disagreed. After a year of working together, Ryden and Norris parted. Most likely it was not amicable.

The *Minneapolis Star* published an article in June of 1959 in which Ryden is quoted as saying, "It's true, that I've been fired by most of the best companies—in fact, almost every place I've worked." The article listed his previous employers—Northwestern National Bank, Engineering Research Associates, Minnesota Paints, Inc., Minneapolis Honeywell, McGill-Warner-Farnham, and Control Data—and corrected him by saying that "not all of these fired Ryden."

In checking the places where Ryden *had been* fired, the reporter came up with various explanations from those who worked with him.

"Ryden is a bright guy, but he's a lone wolf—he's not a big organization type man."

"Ryden is a fast mover and has a million ideas, but he doesn't get along with other people—he can't play on any team."

"He's just a promoter—look at his record."

When asked about Ryden's departure from Control Data, Bill Norris said, "Mr. Ryden and I . . . he left under unusual circumstances and I just won't discuss it. NO SIR! N! O!"

Arnold (Bud) Ryden

In any other period, Ryden might then have disappeared into the backwaters of business and been forgotten. But this was not an ordinary time. The local stock market was heating up and someone with Ryden's unusual abilities could easily get financing and create an opportunity. He had his admirers. An earlier associate said, "We got along well. He certainly wasn't the repugnant type. We were surprised to hear about his troubles in other jobs." Someone who worked for him said that Ryden and Norris had a perfectly friendly agreement to disagree. Thus, there was a broad circle of opinion about the man.

Ryden's parents had moved to Minneapolis from Kansas when he was two years old and apparently he had adjustment troubles even in grade school. He said he knew well the inside of the principal's office at Adams school. At South High he grew into a plumpish, medium-sized youngster with reddish hair and a freckled complexion. He lettered in tennis and sang in the glee club. At the University of Minnesota he took an interest in mathematics and chemical engineering but switched to sociology and graduated with distinction in 1942. "I thought business was predatory," he said later. "I started out to become a social worker... but gradually realized I couldn't make a living as a social worker." He directed the choir at Mount Olivet Lutheran Church for a time, and because of his pleasing baritone voice was hired occasionally to sing at weddings and funerals. When the chance for a scholarship to the Harvard Business School came along, he took it, and graduated with a master's degree in 1943, once again with distinction. He then went into officer candidate school in the army and became a second lieutenant.

In 1958, after he had left Control Data Corporation, Ryden, then thirty-seven years old, together with Willis (Bill) Drake and two others, formed Midwest Technical Development Corp., a closed-end investment company. Bill Drake had helped raise the initial money for Control Data, as described in chapter one, but left within a year, probably because he didn't become director of corporate marketing, a title he expected. He said he "couldn't resist Ryden's offer to form a new company." Ryden explained his own departure in the following terms: "I had to get out on my own... I felt tremendously stymied in every job I had."

The purpose of Midwest Tech was to "develop technical companies and at the same time make a profit for everyone." Ryden believed in sharing the profits with everyone who helped make it happen, and when the stock moved up because of the accelerating enthusiasm of the local market, it was reported at the time that he himself, now rich, was still driving a Chevrolet and wearing $40 suits. "People snipe at me as a manipulator. I'm not," he said. "I hate the fast-buck philosophy. I want to build up businesses solidly."

In February 1959, Ryden led a group of Twin Cities businessmen in the purchase of Telex Products Group, a privately owned company that had been making portable hearing aids and various electronics since 1936. The new group intended to "grow" the Telex business by acquisition. Ryden made a statement that showed his impatient and expansive nature. "Telex now has sales of $5 million a year. We have a timetable of building this to $20 million...within two years. It might be six months, but if I said that, people would say I'm crazy." Telex stock became public as a result of the increased number of owners rather than through a formal public offering, and it went from $17 to $38 in the first six months of 1960. The company soon acquired six other electronics companies and achieved sales of almost $21 million for the fiscal year ending March 31, 1961, meeting Ryden's two-year target. One business executive called Ryden "the greatest financial genius ever to live in the Twin Cities."

In September 1960, *Fortune* magazine printed an article titled "The Egghead Millionaires." Reporters had interviewed more than twenty scientists who made their millions in companies they founded. The thrust of the article was that "the cold war had reshaped the entire job market...expanding the financial rewards that physical scientists could expect." These men were called Space Age millionaires. Most came from the missile industries in southern California; Cape Canaveral, Florida; or the Pentagon area in Washington D.C. Some were from M.I.T. in Boston, the University of California–Berkeley, and Stanford. One of the wealthy scientists said, "Money is 10 cents a bushel. If you dropped me off naked in Miami, I'd have a million in a year and a half." Said another, "There isn't really much you can do with money, except maybe buy other companies. Money doesn't make you any younger, or any happier; you can't jump any higher. Making it is sort of a game, like playing poker." Easy riches can lead to insouciant attitudes like that.

Also listed among the millionaires was Arnold Ryden, a somewhat strange inclusion since he was not an engineer and he was not from the space or missile industries. He was a general businessman specializing in finance, working in Minneapolis. The article reported that Ryden made his first million on Control Data and his second million on Midwest Technical Development Corp., an investment firm that helped finance small scientific firms. Control Data was only three years old then and traded at $46 that summer in the Twin Cities over-the-counter market. Midwest Technical Development Corp., approximately two years old, was trading at $10 a share. Now Ryden was at Telex, his newly acquired company. Bill Norris was not mentioned in the article, nor was Earl Bakken of later Medtronic fame, nor Seymour Cray, nor any other inventive scientists in the Twin Cities area, of whom there were many. Instead, Ryden, the man fired from many jobs, was now featured nationally because, like the others, he was an egghead and a millionaire.

In January 1962, a magazine called *Select* published an article featuring Ryden alone. It was titled "An Ambivalent Look ... at a controversial man, A. J. Ryden." He was described as living in a fine house by the Interlachen Country Club and driving a Lincoln Continental. Although he could easily afford handsome, silken clothes, the article said, because he was impatient he took the first suit on the rack. And, since he was color blind, he and his wife could not agree on oil paintings for their home, but, should they agree, they would doubtless have an expensive original.

"There seems to be some sort of conviction in the business community that a business leader should not be controversial," Bill Drake said, referring to Ryden, his fellow founder at Midwest Technical Development. "Nobody around here is worried about being controversial. We just worry about being irresponsible." And Bill Drake could be counted on to speak the responsible truth. He later went on to found Data Card, a very successful company; and he was a mentor to many start-up companies, invested in them, and served on more than twenty boards, some until his retirement at age eighty. He was appointed a regent at the University of Minnesota, and he also engaged in many acts of quiet philanthropy, such as paying college tuition for those in need. Bill Drake's career suggested integrity. For a time, however, he was entwined with Ryden—not a dishonest person, but an impulsive, controversial one.

Ryden, besides being president of Midwest Technical Development Corp. and president of Telex (this was before the inflation of titles; the term CEO was not in use), was also chairman of the board of two other companies: National Semiconductor Corp. and International Properties, Inc.; and a board member of three more companies: Midwest Instruments Inc., Washington Scientific Industries, and International Finance Corporation. These businesses had little in common, but some of them were new underwritings in the Twin Cities market and the presence of Ryden on their board gave them a certain golden aura. He was glad to oblige as his expansive attention seemed to show an interest in everything. "He has the rare ability," said the president of Gustavus Adolphus College, where Ryden was a trustee, "to detect the issue in any matter under discussion, to focus on it, and dispose of it, with a minimum of discussion." In another instance, when asked by one of his executives if the corporation had enough cash, Ryden responded, "Cash is no problem. If you can make production go, cash is the least problem. I can't conceive of cash being a problem." A Ryden supporter said, "Telex sales were four and a half million in 1959. Bud [Ryden] brought them up to $20 million this year and he's predicting $100 million by 1965"; this, from a hearing aid company he was trying to transform. Obviously Ryden knew how to promote, and his ambitions were real. Repeated success at raising money, plus national recognition, can create such confidence.

Norm Terwilliger, head of the Teachers' Retirement Fund, said, "Sure, some people don't like him [Ryden]. He's a doer. He's done a great deal for the community and the state in starting these new businesses. They mean new jobs, new payrolls...." Ed Howard of Piper Jaffray and Hopwood said, "Ryden is a very high grade man with very high grade moral and business ethics."

Someone who'd been his adversary said, "He can't make small talk. I feel sorry for him; he must be a lonely man. But he's certainly capable in a business situation." Added another, "He's not an agreeable guy to work with. But it's been the most stimulating association I've ever had."

Summing up, maybe it could be said that Arnold Ryden became a social worker after all, lifting many people to a better life. But this was largely because he seemed to leave a trail of money wherever he went during that opportune time. He exemplified the easy success of those heady days.

But markets have a tide of their own. In the broadening swell they carry most everybody until at some point the momentum is unable to sustain itself, and then the breakdown comes, often with equal force. The Twin Cities local stock market hit high tide in 1961. The national stock market was on a buoyant run also, and throughout the country all stocks were ripe for a correction.

It came in the spring of 1962. The Dow Jones Industrial Average fell 26 percent. The Twin Cities local over-the-counter stocks fell even more. In three months from March to June of that year, Control Data and Midwest Tech dropped in half. Economics Labs, SuperValu, and Schjeldahl each dropped 40 percent or more. The new issue market went cold, and with it Arnold Ryden's seemingly limitless supply of cash.

By August 1962, the forty-one-year-old Ryden was already referred to as the ex-chairman of Midwest Technical

Development Corp. The company was being investigated by the Securities and Exchange Commission for "gross mismanagement and abuse of trust." At issue was the sale of Telex stock, both by the company and by officers and directors personally. They shouldn't have been trading in the stock personally, the SEC said, and they should turn over their profits to Midwest Tech. Telex stock had since fallen 92 percent. In his defense, Ryden said, "If I had known the market would fall, Boy! Would I have sold more! I own a higher percentage of my initial holding in Telex than Midwest." Nevertheless, the directors of the company removed him as chairman, and the Telex board also asked him to resign. He refused, saying he had a hand in building Telex, that he was "hurt" and "mad," and wouldn't resign "under fire." At its peak, his personal fortune was estimated at $4 million, and during his run of five years he had been responsible for raising over $10 million of company capital. "He's a trouble maker," Norris said. And the Twin Cities public heard little about him after that.

Jim Secord remembers the market collapse well. A graduate of Washburn High School in 1954, he went to work for Northern Pump/Northern Ordnance in the drafting department designing battleship guns. He had no formal college education—his father had been president of the pipefitters union but was a voracious reader. Jim, in like manner, educated himself through night courses at the University of Minnesota and also by joining the elite Great Books study program from the University of Chicago. In 1955, he became a draftsman for Minneapolis Honeywell Co., and a year later he joined Magnetic Controls, a company newly formed by former Honeywell employees. He became a salesman. In 1960, Magnetic Controls went public, and by 1962, Secord had stock worth $16,000, a splendid sum for a young man and a fund of opportunity. It was enough, he thought, to travel the world for a year. He quit the company and went to Europe, traveling by rail for six months to all the best places on the continent, eating well, drinking the local beer, and taking in the cultural treasures. His plan was to travel next to Greece, leave on a "tramp steamer" (a freighter making local deliveries), head east, and make his way to South America and then home. There was no hurry; he was seeing the world. As he was about to take the next step, however, Secord received a message from his father. "You don't have money anymore because the market collapsed."

Secord went back to America. "I had left verbal instructions with my broker to take action if necessary," he said, "but he was no longer in business." Magnetic Controls stock dropped from $15 the previous year to $2 in the summer of 1962. (It went even lower in the next few years, but eventually combined with a company named ADC Telecom, adopted that name, and experienced a second flush of success many years later.) For Secord, however, the enduring memory was that summer in 1962 when his wanderlust plans ended abruptly and he was forced to come home and go back to work.

The book, *Moneymaking in the Twin Cities Local Over-the-Counter Marketplace,* referred to the year 1962 as the "infamous bust in the Twin Cities local OTC market." Continental Securities published a report saying that "the year will go down in stock market history as probably the most momentous in the last three decades. The spring drop...was the sharpest since the early 1930s."

4

Popular Delusions and Prosperity

The avaricious frenzy that took place in the Twin Cities over-the-counter market during the early 1960s, although unusual, was not unprecedented. In the early 1950s there was a bubble of excitement based on uranium and nuclear power in Utah, with up to five hundred stocks trading on the Salt Lake City Penny Stock Exchange. Three hundred years earlier, Holland was beguiled with Tulipmania—just one of the many examples featured in a book called *Extraordinary Popular Delusions and the Madness of Crowds* written by Charles Mackay in 1841. The book remains a classic even today. Enthusiasm is contagious and so are worry and fear. Such cycles come from the almost unavoidable fervor of human behavior. When riches arrive unexpectedly, people get swept up in the mood of the moment. They anticipate more of the same. And when conditions turn bad, it's hard to recognize and take advantage of opportunities in the pervasive gloom. In the 1930s, for instance—a decade of economic quicksand and investor wariness—sales at Minnesota Mining and Manufacturing Co. tripled. They doubled again during World War II. Surely someone noticed.

Bernard Baruch, the legendary American speculator of the late 1800s and early 1900s (he may have been the Warren Buffett of his time, but he called himself a speculator rather than an investor) was one of many influenced by Mackay's book on crowd behavior. It helped him recognize unsustainable speculation and sell his stocks before it was too late. Many times he missed an opportunity to make even more, he said, but in doing so he also missed an opportunity to go broke, something he had seen happen to others. After gaining personal wealth on Wall Street, Baruch became chairman of the War Industries Board during World War I, and later an advisor to President Woodrow Wilson at the Paris Peace Conference.

In the context of historic bubbles, the Twin Cities stock excitement of the early 1960s might be considered a mere footnote or a minor example of exaggerated crowd behavior. Then again, maybe not. Company after company came public at $1 a share, each promising its own success. When the quick dollars were no longer available, prices began to drop, and investors, one by one, quietly cashed in their holdings. Conversation turned to other things. Some investors, of course, having been unsure why they bought in the first place, held on, equally unsure about their next move. Sometimes, their inaction turned out to be the right decision.

The collapse in mid-1962 could easily have been the end of the bullish excess, but that was not the case. Too many good things were happening. A new economy was developing that would rival and eventually surpass lumber and mining in economic impact on the state of Minnesota. For example, the following stocks that contributed to the area's reputation for electronics came public during that period:

Control Data	Data Display
Baker Engineering	Flo Ttronics
Medtronics	Transistor Electronics
Magnetic Controls	

Interestingly, the number of financial concerns coming public was even larger (abbreviated as shown in the local paper):

Great Northern Ins.	Murphy Finance
Federal Ins.	First Midwest
Amer. Prem. Ins.	Imperial Fin. Services
No. Central Cos.	Dial Fin.

Gen Life of Wisc.	International Fin.
Minn. Cap.	Search Invest.
United Invest.	Equity Cap.
Marine Cap.	No. Am. Life and Cas.

And there were two other companies worth noting because they went on to significant success: Possis Machine, producer of custom-designed automation equipment; and Tonka Toys, a producer of metal trucks and construction toys for children.

The steep downward adjustment in stock prices that happened so quickly in 1962 came to be seen by many as a market correction, rather than a fundamental economic change. Weak companies disappeared, but the valuations of others simply dropped to more reasonable levels. *Perhaps they're worth a second look*, the public seemed to decide. Control Data was still selling computers, and many investors had made a lot of money in the stock. There was continuing hope. In its analysis and forecast for 1963, Continental Securities recommended ten Upper Midwest over-the-counter stocks whose prices, with one exception, "were down at least 30 percent" from the bull market high, and whose "earnings were increasing 14 to 60 percent" from a year earlier. "Select two, three, or five from among these offerings," they recommended:

Control Data	Tonka Toys
Schjeldahl	Rosemount Engineering
LaMaur	Scientific Computers
Pako	North Central Co.
Telex	Mammoth Industries

It was a list of solid companies.

Only ten new stocks were listed in the newspaper in 1963, down from ninety-six two years earlier. It had become far more difficult to get public financing for an untested venture at the enticing price of $1 per share.

Stock manias are sometimes followed by disillusion, even anger and retribution. There was little of that in the Twin Cities in 1962, although some firms and individuals did receive regulatory discipline. Executives from two local brokerage firms active at the time declined to be interviewed for this book. One said, "I don't want to relive those times," and the other felt he had not been treated fairly by regulating authorities and didn't want to participate in any further discussions.

Rather than expiring in a final gasp, the local market then entered a new cycle that would rise to another peak six years later. It would continue the benefits of the former cycle, but not its gullibility. Better companies were brought public, usually at higher prices. There was an impression of substance and quality. And the public played on, not out of foolishness, but because they forsaw continuing opportunity.

Economic bubbles are not all air. They start from real evidence. The value of Control Data's stock increased a hundred-fold in a short period of time. Investors seek out analogous possibilities and some of them turn out to be genuine.

In 2007, Daniel Gross wrote a book called *Pop! Why Bubbles are Great for the Economy*, exploring the phenomenon. He explained his theme as follows: "Time and again American investors, seduced by the lures of quick money, new technologies, and excessive optimism, have shown a tendency to get carried away. Time and again they have appeared foolish when the bubble bursts. But every bubble has a golden lining."

Gross gives examples. "Investors in railroads were frequently left poorer for their troubles...but the entire population grew richer [from] the powerful new commercial and consumer platform." There were 9,000 miles of rail in 1850 and 30,000 miles ten years later. By 1869, a golden spike in Utah connected the country from one ocean to the other. Andrew Carnegie built his steel empire in the late 1870s, feeding the huge appetite of the railroads. By 1885, there were five trunk lines from New York to Chicago, three of them near bankruptcy. A decade later, a financial panic created a recession lasting four years, and about a quarter of the nation's railroad industry fell into receivership. Nevertheless, because of the extensive railroad construction, a housewife in Nebraska could leaf through a catalog, place an order, and receive her goods by rail. It was as though she were not living in loneliness on the western prairie, but instead window shopping on the streets of Chicago. Sears and Montgomery Ward both grew very large on this merchandising approach. Meanwhile, the woman's husband used the railroad to ship his hogs to the Omaha or Chicago stockyards on the return trip.

In 1846, there were 40 telegraph miles in the United States. Two years later there were 2,000 telegraph miles. And four years later there were 23,000, with 10,000 more under construction. The first line from New York to Boston charged 50 cents for 10 words or less, and 3 cents for additional words. Rate wars soon reduced the charges to a penny a word and this greatly helped the development of newspapers (the AP wire service), railroads (timetables), and other large-scale industrial activity. Once the Civil War was over, Western Union consolidated many struggling telegraph companies and became dominant in the industry. The Chicago Board of Trade was set up in 1867, helping to equalize commodity prices in many places across the country, and that in turn led to the first stock ticker machine in 1871. Thus, in twenty-five years, the country went from receiving information overland at the speed of a horse to near-instant communication traveling at the speed of light. A grain elevator in Kansas could check the price of wheat in Chicago, Minneapolis, or Buffalo, New York, and also get paid promptly as Western Union began to offer the additional service of wiring money by telegraph. Investors may have had a rough ride but the general public enjoyed great benefit.

In his book, Gross also cites the recent example of fiber optic communication and the creation of the Internet. That "dot-com" bubble financed the installation of communication webs to many distant locations both here and abroad. Some who invested lost money, but the networks facilitated a global marketplace that many of us now use and enjoy.

"Given the long-term benefits they can produce, bubbles shouldn't be feared so much as regarded with concern and respected" the author says. RESPECTED, that's a strong admonition, but the evidence he cites supports it.

Though unmentioned in the book, the Twin Cities local over-the-counter market in the 1960s had a similar effect. According to a report at the time, there were 117 electronics firms in the area at the end of November 1960, with at least 30 new companies formed in the subsequent six months (not all of which raised money in the public market.) Total employment in the new industry was reported to be 40,000. This number most likely included Minneapolis Honeywell, Univac, Control Data, and possibly the IBM plant in Rochester, but it surely also included the dozens of new start-up companies. An industry was being defined. Control Data had 690 employees on June 30, 1960, and almost double that a year later.

In a demonstration of how unique and embracing the Twin Cities new issues market was, *Mid-American Investor* magazine, in February 1963, published a comparison of local stocks to those in other markets throughout the country's midsection. The Minneapolis/St. Paul market had 105 local listings. Even though many local stocks had by then merged, gone out of business, or gone to the national over-the-counter market, this number was almost four times larger than the comparative number in the huge Chicago metropolitan area, which stood at 28; and more than three times Milwaukee's total at 32. St. Louis and Kansas City had a total of 59, which, if split equally, put them in the range of 30 each, about the same as Milwaukee and Chicago. Houston, at 57, came in higher, but still short of the Twin Cities by almost half. Dallas and Denver were unmentioned. The burst of local financing in the Twin Cities was unusual in its size and impact.

Caught in the sharp local market correction of mid-1962, Possis Machine dropped from 25 (bid) to 15 in three months, but didn't stay there long. On December 31 of the same year, it closed at 47. Perhaps that was why it was not mentioned in Continental Securities' stock recommendations for 1963; it had already tripled off its low. But that rebound wasn't just a whoosh of rising air, because Possis was a real company capitalizing on the new demand for automation. Its founder, Zinon C. (Chris) Possis, was a very creative man. "We translate ideas into machines," he said at the time.

During that same period, Northwest Airlines, although not trading in the local over-the-counter market, went from approximately 12 at its low in 1962, split 2 for 1 in 1964, split 2 for 1 again in 1966, and hit a high that year of 60, an increase of 20 times in four years. Certainly such propulsion caught the attention of serious investors in the Twin Cities and became part of their conversation.

Another company, Tonka Toys, also drew investor interest. Located in Mound, twenty miles west of Minneapolis, it was called the world's largest maker of metal model trucks and other vehicles. Production in 1961 came to nearly three million units, almost as many units as General Motors produced that year. Yes, General Motors. Each toy, however, sold at retail for $2 to $13. In July of 1961, the company offered 155,000 shares in its initial public offering, priced at $12 a share, and soon thereafter began paying a dividend of 50 cents a share, something quite unusual for a new offering in the local over-the-counter market.

With the money raised from the sale of new stock, the company planned to increase plant capacity and add jeeps to its line. Sales were then expected to approach $10 million annually. By early 1962, the stock hit 46, but along with others caught in the market correction that spring, it fell to 23. It went on to better prices in following years, for this was a real company making real products, being sold to real customers, and it had the added bonus of a cash dividend.

Stock in these three companies differed from most of the earlier offerings in the Twin Cities market in that they were not just a promising idea. They had activities that could be measured and valued, and they produced tangible returns significant enough to keep the investment fever alive. But the street bazaar of dollar stocks in every flavor was over.

Of the three companies, Possis was perhaps most typical of new economic growth made possible by the Twin Cities local market. Its success grew out of the ambition of a highly motivated, creative person, wanting to build something of value—not just a product, but a company capable of continuing success.

Born in St. Paul on June 22, 1923, Zinon C. "Chris" Possis was the oldest of three sons. His parents were Greek immigrants, and he was not much exposed to English until kindergarten. Nevertheless, he learned quickly and was allowed to skip the second grade. His physical development did not come so quickly, however. Only 4 feet, 6 inches tall as a freshman at Mechanic Arts High School, he said later, "The fact that I was small and smart drove me." That determination must have had an effect. He grew to 5 feet, 6 inches and graduated as class valedictorian in January 1941. He attended Dunwoody Institute for six months and then moved on to mechanical engineering at the University of Minnesota, but only briefly. In 1942, he left school and joined the U.S. Army as a paratrooper and became a demolitions expert sent to the Pacific. At the age of twenty-two

Chris Possis (right) and a colleague examine a part.

he became the army's youngest division staff officer and was recognized with a commendation medal from General Douglas MacArthur.

When the war ended, Possis, like so many others, joined the crowd of returning veterans and reentered the University of Minnesota, graduating in 1947 with a degree in mechanical engineering. As a student, he had found work at a company called Shopmaster, and upon graduation joined them full-time. He soon became chief engineer and supervisor of some three hundred employees. His abilities now recognized by management, he felt entitled to buy a piece of ownership in the company, but his request to do so was rejected. So, in 1952, the year of his twenty-ninth birthday, Possis left the security of corporate life and formed his own company. He would be a sole proprietor, renting space on the third floor of a building on Hennepin Avenue in Minneapolis, seeking customers who would pay for his unusual technical abilities.

"He had a touch of genius," said attorney Joe Walters, who was his neighbor at that location and knew him even before. In 1956, Possis incorporated the proprietorship under the name Possis Machine Co. Much of the company's work involved automating industrial equipment, and much of it represented one-time projects. One of the company's inventions was a machine that assembled the tiny parts of a wrist watch at the rate of seventy-five parts per minute, replacing employees working with tweezers and magnifying glasses whose eyes needed a rest every two hours. The machine illustrated the company's unique talent in very specialized areas. Possis Machine claimed to be the country's "second largest manufacturer of automatic machines related to the production of electrical motor parts" and hoped in three years "to be the biggest."

Sales grew every year and the company was profitable.

1957 total sales $45,000

1958 total sales $153,000

1959 total sales $367,000

1960 total sales $595,000, net profit after tax: $43,000

Coincidentally, the local market had become very active by this time and underwriters were looking for a compnay with a record like this. Various associates described Chris Possis as "creative and intense, hard-working and honest," characteristics an investor should like. Sales were approaching $600,000, and the company had a record of consistent profitability; therefore, an initial stock offering could be priced as a successful business, not a speculation at $1 a share. Possis decided to offer 40,000 shares for approximately 30 percent ownership, with 100,000 existing shares retained by management.

On August 22, 1960, Craig-Hallum Kinnard, Inc. brought the company public at $7.50 a share. This valued the total company at approximately $1 million, an amount seemingly fair to investors because of the company's likely future prospects, and fair to the founder and his employees who had brought the company from nothing to its current status. Before year end, the stock traded at 10. Construction began on a 20,000-square-foot plant on a four-acre site in the Minneapolis suburb of Golden Valley. A year later, Possis had eighty-two employees, most of them tool and die makers or skilled machinists. Demand for the company's products was so strong that employees were working an average of fifteen hours a week overtime. Sales hit approximately $800,000 in 1961 and were expected to reach a million the following year. "We're turning down work," Possis said. The company had developed a taping machine that had revolutionary characteristics, and when Minnesota Mining and Manufacturing Co. expressed interest, it became the company's first mass-produced item. With only 40,000 shares in the public market, it didn't take much buying to cause the stock to triple in price in the last half of 1962. To help increase liquidity, the company split the stock 2 for 1 in October 1963. In 1965 the stock was listed on the national over-the-counter market.

Possis Machine had also developed a process that many believed would revolutionize film packaging. Eastman Kodak had paid them $350,000 for the patent rights, intending to develop a machine of its own using the technology. Kodak agreed to pay Possis a royalty of $10,000 per machine per year. It looked like an oncoming surge of cash. Some projected that the company could show earnings of $25 to $30 a share.

Speculators rushed to buy the stock, and the price went crazy, moving as much as 40 points in one day. No doubt market-makers were squeezed when they sold "short," and this contributed to the startling acceleration as they tried to buy shares to cover their position. In April 1966, the stock hit a peak of 245. Chris Possis was relieved there wasn't much trading at top prices. "It would have really hurt me to see a lot of buyers at that level, but thank goodness there weren't," he said. Those who bought at the bottom in 1962 and were lucky enough to stay aboard for the ride to the top four years later saw the value of their holdings increase nearly thirty-fold. Chris Possis's personal holdings were worth over $15 million. Later, he commented on the irony of that run, saying it occurred in the only year the company did not make a profit. Revenues were $4.1 million, and the company lost $1.22 per share. A year later the stock fell 85 percent.

Although Eastman Kodak spent three years working at it, the packaging machine was too expensive and suffered too many breakdowns. In Chris Possis's opinion, it was "poorly designed." Eastman Kodak eventually dropped the project and the patent rights reverted back to Possis Machine.

A few years later, Chris Possis's two younger brothers, Emmel and Milton, started a securities firm dealing in Twin Cities over-the-counter stocks. Chris insisted they not get involved in the stock of Possis Machine because of a possible conflict of interest.

In the decade that followed, the company began to focus on medical products. Perhaps this new direction was a coincidence, perhaps not, but after the birth of their daughter, Chris Possis's wife became sick, "beginning with headaches, backaches, and depression," as reported in an extensive profile of the man and his business in the *St. Paul Pioneer Press*, January 16, 1984. The article described Mary's ordeal in detail.

> *During a 15 year time span, Mary Possis was hospitalized for six years, underwent brain surgery three times and had three spinal operations, and was given a wide variety of medication for an illness the doctors could not diagnose.*
>
> *[In 1976] a friend suggested that her problems might be linked to chemical [pill] dependency. She entered St. Mary's chemical dependency program and seven weeks later "walked out a new woman," Chris Possis said. "Since the day she got out, she has not touched one pill."*

But Possis says he does not blame the doctors who treated his wife for compounding her problems by prescribing more drugs.

"For what she's been through, it's amazing she's alive," Possis said. "She's fearless, courageous and just a beautiful person inside and out. I love her more now than I ever have."

In comparison to this trauma, it can be said that business setbacks are nothing, and that the man's forgiving attitude bordered on sainthood. But his wife's struggles no doubt affected his work. And during the period 1975–1977, the company reported losses for three years in a row. In the middle of that period, fifty-two of the company's seventy-six workers went on strike, and Chris Possis, ever the man of strict standards, said the company could not afford to increase wages. The strike lasted fifty-nine weeks—certainly a test of patience and maybe even survival for all parties. The stock dropped below a dollar a share. Cumulatively, company losses over the three-year period totaled almost a million dollars, and Possis used $75,000 of his own personal savings to meet payroll.

When the strike was settled, none of the striking employees returned to the company. It became profitable again, but by then, Possis had directed his attention to medical products. In 1972, he met with Dr. Demetre Nicoloff, a University of Minnesota cardiovascular surgeon, with the idea of developing an improved mechanical heart valve. This eventually proved productive, but, as in most things, there were complicating factors. Nicoloff had developed the valve in his own personal company and then transferred ownership to Possis Machine. Rights to the valve were sold to St. Jude Medical in return for royalties. A legal dispute followed.

"[Possis is] not inclined to delegate responsibility," said Alan Ruvelson, then president of First Midwest Corp. "And he's very principled—a straight arrow. He believes what he says and has a high level of confidence in what he says."

It seems that Chris Possis's uncompromising behavior served him well in matters of technical development, but not in those areas of business calling for a measure of flexibility. Possis contracted cancer, was ill for a number of years, and died on October 16, 1993, at the age of seventy. His company continued on successfully until it was acquired in 2008. Today there is a plaque in the lobby of North Memorial Hospital on the edge of Minneapolis dedicated to "notable people" that honors his service on the board of trustees for twenty-seven years and cites him as the inventor of the artificial heart valve. The local stock market played a role in that discovery, helping to fund his work.

5

Cinderella Grows Up

Control Data's remarkable rise from $1 to $100 a share in four short years was a Cinderella story, worthy of the local headines it received. But what seemed like a milestone was in fact only an interim stop. Across the country, many were just learning about this extraordinary little company located in Minnesota, a state better known for agriculture, mining, and recreational fishing than for hi-tech wizardry. They were learning that the company was able to compete against IBM on a profitable basis when half a dozen larger companies were unable to do so.

Investors naturally looked for reasons to explain Control Data's unusual success. An obvious reason may have been that other companies didn't have an employee comparable to Seymour Cray. Another was that much of the company's business was with the U.S. government, which purchased rather than leased their machines, while the competition sold to commercial customers who almost always leased their machines. Leasing delayed revenues and made it difficult to offset the heavy up-front expenses necessary to get an order. A third reason was that Control Data machines were sold at the very high end of the market—one of its systems sold to the U.S. Atomic Energy Commission for $7 million—which, of course, created revenue sufficient to offset the cost of building the machine, plus the cost of ongoing development, front end marketing, and initial installation and service expense. In 1961, the company's profit margin before tax was 11 percent, not an exceptional level of profitability but certainly a comfortable one given the conservative accounting. In some ways it was like explaining how Switzerland managed to stay at peace when all it neighbors were at war. The terrain helped.

In September 1961, Control Data stock split 3 for 1 and six months later hit $50 a share, which meant, of course, that it had advanced 50 percent from its $100-a-share milestone four months earlier. The company was now pursuing foreign markets by opening offices in Australia, Canada, Switzerland, The Netherlands, Sweden, West Germany, and France. Such sales expansion called for production capacity. A new plant had been built in a southwest suburban location to produce computer peripheral equipment—that is, card readers, printers, tape handlers, and related accessories; and eventually disc drives. Plans were also being considered to build a large computer manufacturing plant in Arden Hills, a suburb north of St. Paul. And Seymour Cray wanted a quiet place to work, so an engineering lab had been built in Chippewa Falls, Wisconsin, his home town, ninety miles east of the Twin Cities. His development staff would move there with him to work on his next supercomputer, the sixty-six hundred (6600). The building was outside of town in the woods along the Chippewa River, far enough away from civilization so Seymour could concentrate without interruption. There were reports that he initially had the telephone installed in a phone booth outside the building, or on a tree, but Les Davis, a long-time colleague of Seymour's, says that this was just part of the Seymour Legend. Everyone who needed a phone had one.

But when markets are gripped by fear, as they were in 1962, a company's formerly promising expansion suddenly begins to look like additional expense that might not be covered by revenue. Investors worry and stock prices begin to fall. The recently split Control Data stock dropped in half, and on June 30 hit $25 a share. This, in retrospect, was another golden opportunity to buy for those willing to bet on continuing success and willing to hang on for a bumpy ride. In the spring of 1963, the stock recovered to $40 and was accepted for listing on the New York Stock Exchange—a prestigious move for a company not quite six years old.

The *New York Times* published an article on August 30, 1963, under the headline, "Control Data Corp. Stock Continues to Soar," as though it were now a familiar matter to their readers. The tag line said, "250-Fold Increase Is Shown by Shares in Six-Year Run." The stock closed at 83¾, and was the "16th most active issue crossing the ticker tape that day." It was stunning how far Control Data had come, and how fast.

The *Times* article stated that no other issue listed on the New York Stock Exchange could match Control Data's phenomenal rise during that period, calling it "one of the most fabulous six year runs in recent history." It described the company's beginnings in the over-the-counter market "when small unknown companies turned the investment public goggle-eyed with excitement. Many later fell by the wayside…but Control Data, after weathering the general market break last year, set the stage for another long advance. Fueling the gains were sharp increases in fiscal 1963 sales and earnings." It was called a "classic Horatio Alger story at the corporate level."

Norris was invited to speak to the New York Society of Security Analysts in September, the second time he had been invited. Analysts wanted to better understand the company, assess its possibilities, and size up its leader. They wanted to learn what this company had that others didn't have. "Brains," Norris said in a later interview—that was the difference. Then he added clarification. "Control Data and IBM are the only two companies in the computer field making money because they're the only two just in that business. In computers the decisions are pretty big and must be made promptly. In big multidivision companies like RCA, Sperry Rand, Honeywell, or General

Electric, top management is engrossed in many things and is not knowledgeable about the problems in the computer division." The remarks were certainly tinted with his feelings about earlier work at the Univac Division of Sperry Rand. He used the phrase "absentee management" to describe his struggling competitors.

The *Minneapolis Star* observed that the young company was selling at "a price-earnings ratio of 100 or more, highest on the exchange," and cited in comparison two other growth stocks against which the company was sometimes compared: Polaroid, selling at 80 times earnings, and Xerox, selling at 70 times earnings. IBM's ratio was approximately 41 times earnings. A ratio of 20 times earnings was considered "the norm." At such extended levels, investors were undoubtedly expecting continual growth on into the indefinite future. Skeptics, on the other hand, said that such prices anticipated not only the future, but the hereafter. Computers, instant cameras, and copying convenience on regular paper were the latest technologies, and Control Data was leading the field, selling at the highest multiple of earnings.

This exalted status was much deserved. When the supercomputer Seymour Cray had been working on, the 6600, was announced at a press conference, its computational capabilities clearly outclassed the competition. IBM's president, Thomas Watson Jr., wanted answers. An internal memo, dated August 23, 1963, that expressed his feelings and acknowledged the superiority of the machine, found its way into the general press.

> *Last week CDC had a press conference during which they officially announced their 6600 System. I understand that in the laboratory developing this system there are only 34 people including the janitor... Contrasting this modest effort with our own vast development activities, I fail to understand why we have lost our industry leadership by letting someone else offer the world's most powerful computer... I think top priority should be*

given to a discussion as to what we are doing wrong and how we should go about changing it immediately.

Perhaps he knew, but couldn't acknowledge, that a work of genius, by definition, is difficult to duplicate. But not impossible. The "brains" Norris referred to earlier surely had to include Seymour Cray. To receive such product acknowledgement from the company's principal competitor in a moment of candor was truly extraordinary.

Something else Norris now felt sure about was that his company had currency that should be put to use; that is, currency in the form of stock selling at a very rich multiple of earnings. He began to broaden his base of business by making acquisitions and paying for them with stock. "Chinese money," it was called, meaning it had inflated value. (China had an undeveloped economy at the time.) If the acquisitions had no earnings and weren't too large, they wouldn't much dilute the earnings per share of a company selling over 100 times earnings; alternately, if the acquisition had earnings and was bought for a lesser multiple, it would increase the resultant earnings per share. As in other matters, once he decided, Norris pursued the activity vigorously. In some ways, this strategy was what Arnold Ryden had recommended for the company in its early years. Now it was being done according to Norris's timetable and within the discipline of building on the base of existing business. The fundamentals were firmly in place. The company's computer business was thriving, and new capacity had been created to build peripheral products that were also contributing to revenue and earnings. It was time to buy new customer connections, service reach, and technical talent. He had the money—er, Chinese money in the form of high-priced stock—to do so.

In calendar year 1963, the company acquired five businesses: the Computer Division of Bendix Corporation; Digigraphic Systems of Itek; Control Systems Division of Daystrom; Becks

Inc.; and Electrofact N.V. in The Netherlands. In the following year seven more businesses were acquired: the Stromberg Transactor business of General Time; Rabinow Engineering; Holley Computer Products; Bridge Inc.; Computer Laboratories, Inc; Adcomp Corporation; and TRG, Inc. Most of these brought new customers to Control Data; some brought service support or new products, and others brought unique talent in the areas of optical character recognition and laser technology.

In July 1964, the stock split 3 for 2, meaning an early shareholder now held 4.5 shares for every original share. A month later, *Time* magazine called the company "the poor man's IBM" and added that the term was used lightly as the company's fast growth had made it a favorite "glamour stock" on Wall Street. And in October of that year, *Dun's Review*, a national business magazine, published the article "What Makes A Growth Company?" and selected thirteen in the country that truly fit that description, a daunting effort, they stated, since there were 1,242 companies on the New York Stock Exchange, 945 on the American Stock Exchange, and 25,000 others in the U.S. with assets of over $1 million. Here were the "lucky thirteen," as they called them.

International Business Machines Corp.

Minnesota Mining and Manufacturing Co.

Xerox Corp.

Litton Industries

Avon Products

Control Data Corp.

E. J. Korvette

G. D. Searle & Co.

Union Bank of Los Angeles

The Singer Co.

Northern Illinois Gas Co.

Purex Corp.

Bristol Myers Co.

In calander year 1965, intending to meet the industry need for additional programmers and technical personnel, Control Data began a for-profit training division called the Control Data Institutes. During the same year another eight companies were acquired: Data Display; Datatrol; Computech; Glenn W. Perkins Associates; General Precision commercial computer operations; Howard Research; Computing Devices of Canada computer systems division; and Waltek, an assembly operation in Hong Kong. There could be no doubt now about Norris's ambition. He was aiming to get big, and to build a durable foundation in all aspects of the business so that the company could survive against IBM, still the dominant force in the industry.

But as this corporate express was gathering mass, the engine began to make strange noises and cause the vehicle to shake. The company had come out with a new line of computers aimed at general business customers in the mid-price range (the 3000 series). Customers at that level preferred to lease rather than purchase, a condition similar to that plaguing other competitors in the industry struggling to make a profit, and it put a strain on Control Data's earnings. To help improve its profit picture, the company changed the very conservative accounting policies of previous years and took a more liberal, but still generally accepted, approach. The depreciable life on its leased computers remained at four years, shorter than IBM at an average of five years and Honeywell at six years, but the heavy front end charge, at 50 percent of product cost, was reduced by half and spread evenly over the four-year life. The company also deferred some marketing costs, spreading their write-off over the life of a lease but not to exceed three years. These changes meant that the alternative cash flow number highlighted in earlier years was less significant. Investors probably felt that a characteristic that had bolstered their trust in the value of the stock had now been changed. The new mid-range computers competed more directly with IBM and others in the industry. There were rumors the company would have to cut its prices to meet competition. In addition, the company was having some technical problems with its supercomputers. Overall, the torrid rate of growth that had so characterized the company until that time was seriously threatened. Revenues for the year ended June 30, 1965, were $160 million, up from $121 million the year before, a disappointment from expectations, and earnings per share advanced only 16 percent. More telling, however, was that without the accounting change, earnings per share would have fallen from 84 cents a share the previous year (adjusted for acquisitions) to 60 cents a share, the first drop in eight years of business. As investors became aware of these diminished results, the stock dropped 50 percent from mid-1964 to mid-1965 and was now selling in the low 30s.

The *Minneapolis Tribune* ran a headline question: "Control Data a Disappointment?" and quoted Norris saying, "When I say the outlook is very good for substantial continued growth in sales and earnings, I am speaking more in the long-term sense—in other words, averaging out the peaks and valleys, the highs and the lows."

Dick Jennison, the analyst who originally recommended the stock to Wall Street institutions, put out a short memo explaining the company's sharp decline and concluded, "We do not feel that the recent developments alter the company's fundamental strengths for the future. In fact, a basis exists for a positive attitude on the stock."

Forbes magazine followed with an article titled "Is Control Data in Trouble?" In it Norris explained the turn of events. "If a couple of buyers suddenly decide to lease, bang, there go your predictions right out the window," and he declined to predict results for the coming year. When asked if the company had enough capital to continue expanding so rapidly, he responded in a way that sounded like Arnold Ryden a few years earlier, "Capital shortage. Hell, no. That's the easiest part of running a company if it's a good one," and mentioned that the company had a $60 million line of credit. He then berated his critics: "Too many Wall Street analysts don't know their butts from third base. I've always been amazed at Wall Street. I just stand with my mouth open. There's been a lot of superficiality about us all through the years."

Perhaps Norris meant that the stock should never have been priced at over 100 times earnings in the first place, since it created unreasonable expectations; perhaps he meant that analysts too often let mob psychology prevail over serious thinking or that Wall Street was focused on short-term results rather than underlying fundamentals. The remark did not engender a thoughtful discussion and was probably not intended to. Certainly investors form their expectations about a company's future based on performance in its recent past. Such reasoning may be superficial, but Norris himself had financed his expansion on the strength of such expectations. Control Data was now struggling and deserved to be scrutinized closely, which made Norris's remarks sound both defensive and intemperate.

"The company does not expect a good first quarter," Norris said a week later, with earnings "amounting to plus or minus a few cents a share." He predicted results would be better in the second quarter and that the second half would show continued improvement. He stated that a competitor was harassing Control Data with frequent announcements of product and price changes, and that such tactics were causing confusion in the supercomputer marketplace and keeping Control Data's potential customers from making commitments. The unnamed competitor was clearly understood to be IBM. In spite of the difficulties, he ended with an affirmative, "I believe a big future lies ahead for Control Data."

But 1966 was a year of tumult. On February 8, Frank Mullaney resigned, for "personal reasons," of course. He had been with Norris at the inception and held responsibilities at the highest level throughout. He was now forty-three years old and wealthy. "I don't care to discuss the reason for my resignation," he told the *Minneapolis Tribune*, and mentioned his future plans "aren't firm right now." He would remain as a director, however. Two weeks later, Robert Kisch, another of the twelve original employees, also resigned for "personal reasons." Rumors circulated that Seymour Cray was planning to resign, but when asked by *Electronic News*, he denied it. A hotline from "Computing Newsline" said that the explanation they received from "several sources" was that Mullaney and Kisch wanted the company to remain a computer hardware company, while Norris was pursuing acquisitions and diversification into peripheral equipment, training, and other areas. Having acquired so many entities, the company now had more than 11,000 employees and many individual plants to manage. It was being referred to in the press as the *former* "wunderkind of the computer industry."

A few weeks later, the company reported that it had replaced its $60 million credit line with a $120 million revolving credit agreement. The funds would be used to support increasing leased equipment orders, now 60 percent of the company's business versus 25 percent five years earlier. "We've survived the worst," Norris told *Electronic News* in February 1966, "the past two years can best be described as brutal." But it wasn't quite over.

After resigning as an officer in February, Frank Mullaney resigned as a director in July 1966, saying, "I just didn't feel I was a part of that management's team any longer." Seymour Cray also resigned as a director, but stayed on as vice president of research. He was forty-one years old, and in his frank manner stated that his willingness to continue as an employee was out of devotion to his work more than loyalty to Control Data.

Allan Rudell, the company treasurer, also resigned, as did two divisional vice presidents, Ed (Pete) Zimmer, who had been employee number thirteen and was responsible for designing the company's mid-range computer offerings, and Raymond Whitney, vice president of western region sales. Whitney said, "I still think Control Data is one of the best companies going. The assets are fantastic—if only someone could figure out what to do with them." Taken together, the exodus seemed to point to dissatisfaction with the leadership of the president, William Norris. One who left, although unidentified, was more direct in his statement. "I just don't like the way he does business. [He] tries to do everything himself." In a rejoinder concerning the defections, Norris explained his view on the matter: "Jobs change, but sometimes people don't," and discussed the subject further in the annual report of that year. Acknowledging the success of earlier years with the original employees, he then said, "The nature and requirements of management of a large company are quite different from a small one and it became necessary to make some changes. During the course of realignment, we have had a few employees who were either not qualified or did not desire to assume changed positions, and are no longer employed by the company."

The biggest concern was Seymour Cray, of course. He indicated he had been trying to get off the board of directors for two years. "The company at the start was largely technically dominated," he said, "but that is no longer true and so my interest has diminished. I am principally a technical man," and then he added bluntly, "We all have ideas on how to operate, so naturally I don't agree with the president on how the company is running." The remark seemed to suggest that Cray's resignation from the board was perfectly normal, and that his behavior explained his motivation. Or, it could have meant that given his independent personality, he wouldn't have been in agreement with anyone running the company. It was open to interpretation. He also said he was able to operate freely in his research lab and had no complaints about that. Two weeks later, a new quote betrayed his true sentiments. "Years ago, when the company launched into its expansion program, I kind of felt sorry." Sorry, as in sad. It was now clear he did not support the company's expansive ways; he had been a board member, and his sentiments had not prevailed.

Like many creative people, Seymour did not like the administrative details of corporate life. When Norris demanded a written five-year plan from every operation within the company, most gave it serious effort and followed the recommended format, describing the requirements for different courses of action, the method of achieving the chosen course, and so on. Some of the plans ran to twenty pages or more, and included charts and statistical summaries. Seymour, on the other hand, submitted his on a half sheet of paper. His five-year goal: "Build the world's most powerful computer." His one-year goal: "Accomplish one fifth of the above." His method of accomplishment: "Proceed with confidence." Of course, no one else in the company could be so audacious. Seymour was allowed to say, in effect, *trust me, I know what I'm doing*, because he had already delivered many times on such trust. One can envision a sly smile spreading across his face as he submitted the plan.

Cray continued working on a bigger, faster computer, the 7600, at his lab in the Wisconsin woods. For relaxation, he pursued his own unusual interests by building a windsurfing device in the spring…and burning it in the fall to keep roving kids from injuring themselves with it on the winter ice. Some reports said he built a "sailboat" in the spring and burned it in the fall, intending to build a better one the following spring, but colleagues report that this was an embellishment of the Seymour legend. About this time, *Fortune Magazine* said of him, "There is no doubt that, in a field where genius is almost taken for granted, he is a towering figure."

Norris, meanwhile, was described by the *Wall Street Journal* as "tight lipped, austere, and slow to unbend before outsiders." He admitted being shaken by events of the past year, but insisted that some of the problems came from outside the company. "IBM has been out to get us—and you can print that," he said testily. International Business Machines Corp. was reported at the time to have more than 65 percent of the computer business. *Fortune* magazine did a piece entitled "Control Data's Magnificent Fumble" praising the company highly for its bold challenge to IBM's industry dominance, but explaining that this was no easy task and that the stock was "badly battered" because of current problems.

As the fiscal year came to an end in June 1966, the company reported a loss of 38 cents a share, down from a profit of $1.06 the previous year. "Control Data is going through a transition," Norris said. "It is a small company growing into a big company." Since the company earnings missed expectations so badly, some commentators speculated that the bank consortium behind the company's new line of credit had urged a tightening of management and caused the consequent departures. Continental Illinois, the company's lead bank, had, in fact, recommended that George Strichman, a tough, demanding executive from Colt Industries, be put on Control Data's board of directors. The banks wanted to provide more outside oversight. At the first meeting Strichman attended, someone who was there said, "He almost ripped the place apart." Earlier in his career Strichman had been a protégé of Harold Geneen, famous at the time for his authoritarian way and for building IT&T into a dominant business.

Also affecting results in the year just completed were export restrictions put in place by the U.S. government. Over the previous four years, the company had delivered more than $100 million worth of computers to customers outside the United States, and it was now operating in twenty countries. France was the biggest foreign market for Control Data's supercomputers—the company had orders pending for at least five new supercomputers from that country. But for the previous six months, U.S. regulators had refused to grant export licenses on those machines because of restrictions by the nuclear test ban treaty. Although sensitive to the attitude of its government, the company was frustrated in trying to find a satisfactory solution. Further affecting company results were technical issues on some large computers under contract to the Atomic Energy Commission, as a result of which the company had been forced to make penalty payments estimated at nearly $2 million. Other customers were demanding more software to go with their hardware. Reflecting these many problems, the stock hit a low for the year of 23⅝ on October 25.

Eight months later it would roar past 100. "It was like a dam breaking," Norris said.

In March 1967, IBM announced it would stop taking orders for its supercomputer, the so-called "paper machine" that Norris had been criticizing. He had said that IBM's product announcements were premature and were designed to cause confusion among buyers, and that the proposed machine was having serious technical difficulties and was not ready for delivery. With this sudden retreat by the industry colossus, he was now vindicated. Meanwhile, the technical problems on his own machines had been rectified. Customers began placing orders again. After much review, the U.S. government granted export licenses for the machines going to France, a cumulative value of at least $20 million. At the same time, the cost of computer components (mainly transistors) was coming down, allowing some price adjustments. Without the new flow of orders, the company had been caught in a big expansion of marketing and an expensive introduction of new products. It needed revenues to cover those costs, and now they came in a gush.

The *Minneapolis Tribune,* in its Sunday edition on July 30, 1967, published a chart on the price of Control Data stock going from the low 30s in January to approximately 115 in July, and referred to it as Control Data's "stock leap." Indeed, the ascending price almost jumped off the page. The biggest move was in the month of April, right after the IBM announcement concerning its supercomputer, as the stock went from the mid 50s to the mid 90s without pause. In May it caught its breath for two weeks and then resumed its upward sprint.

"The past year has been one of the most successful in the company's history," Norris said in his June 30 year-end report, "as the company continued its worldwide growth." Revenues amounted to $245 million, compared to $167 million the year before. Earnings per share were 98 cents, compared to the previous year's loss of 38 cents. To finance leases, the company had signed a new $175 million revolving credit agreement. Everything seemed in place for renewed prosperity. Since the company was again performing well, the management defections that seemed so telling just a year earlier were forgotten, and so were the accounting changes. Before the end of the calendar year, the stock hit 165. In December, *Newsweek* published an article entitled "The Sweetest Stocks of '67," and number one on the list was Control Data, up 392 percent in the first eleven months of the year. In comparison, the Dow Jones Industrials had risen 11 percent. Two months later, *Fortune* magazine described Norris as "far from formidable, even a bit shy," saying he "smiles gently" even as his business behavior is "go for broke."

Investors again believed in the company's continued success. If earnings could double again, the stock would be selling at eighty-four times next year's earnings. In comparison, IBM was selling at about fifty times current earnings. The market was in love with the idea of growth. Norris's dogged leadership had won out and Control Data had recovered its previous aura. Factoring for splits, investors who bought the stock at the initial offering now had a paper gain 740 times their original investment—an astonishing number for any investment, but especially so within a ten-year period; the investment return was almost triple the increase the *New York Times* had marveled about four years earlier. Such an investor would have felt like he was on the back of a wild bronco at times, having hung on during the early days of tight finances and salary reductions, also during the general market collapse of 1962, and during the swoon in 1965–1966, when company earnings vanished, all to arrive at this moment, not yet a conclusion but a moment of unusual accomplishment.

Control Data's success no doubt raised the overall wealth of both the Twin Cities and the nation. It was successfully competing with IBM, and its future seemed bright. The company was now referred to in the trade press as the undisputed world leader in the very large computer field. Accordingly, its stock had performed like a whippet in a race against overweight mixed breeds and undeveloped mongrels—except for IBM, of course, which was a Great Dane, in a class of its own and a model of growth but less dramatic in comparison because of its size. And in fiscal year 1968, Control Data's earnings came through as expected, better than doubling, to $2.12 a share versus $0.98 a share a year earlier.

One ongoing issue remained. Because two-thirds of Control Data's shipments were now being placed on lease, there was a continual need for new capital. Bank lines of credit had been secured over the years, then increased, and then periodically repaid from the proceeds of additional stock sales. These transactions replaced borrowed funds with new permanent equity capital and reduced the company's balance sheet risk. But it meant going to the public market when the company needed money and selling stock, or debentures convertible into stock—a chancy requirement, as markets sometimes were receptive and sometimes weren't. A young company continually raising money this way was not operating from a position of strength. Competitors such as IBM, with is vastly superior resources, had no such concern. Mindful of their continuing need for financial support, Norris and his then vice president of finance, Harold Hammer, looked for an opportunity to reduce their vulnerability. Hammer, earlier in his career, had been a consultant on Wall Street. He had connections.

Companies were getting bought out in the national markets, sometimes in a friendly fashion and sometimes not. Acts of near-barbarian aggression were taking place. They were not illegal but often were unwelcome by the management of the target company, even though it meant a higher price for the shareholders. After being acquired, many companies lost their mission and suffered jarring changes in employment. The recent history of such buyouts had been ominous. Therefore, those receiving an offer from a suitor they considered not seriously interested in their customers and employees, frantically looked for a friendlier alternative, a "white knight," who would rescue them, respect their years of effort at building a business, and provide an alternative reward for their shareholders. When Commercial Credit Company, a big financial firm located in Baltimore, Maryland, received an unsolicited tender offer for its shares from Loew's Inc., a theater and hotel operator, Commercial Credit said that the offer was designed "to benefit Loew's stockholders at the expense of Commercial Credit shareholders," and the company immediately looked for an alternative. Apparently aware that Control Data had a history of acquiring businesses and helping them thrive, the company approached Control Data about an alternative buy-out offer. Control Data agreed to give it consideration, even though Commercial Credit was nine times larger when measured by assets.

Earnings between the two companies were nearly even, however, and the stock market was much enamored over earnings, and particularly earnings growth. Control Data, of course, was considered a technological leader and, having recovered from its earlier losses, was again considered a true "growth stock." Commercial Credit had none of those characteristics. It had begun business in 1912, helping companies secure credit in ways that traditional banks didn't, such as factoring accounts receivable, asset-based lending, and equipment leasing. It grew to large size and then, comfortable in its position, showed no growth in earnings for the most recent decade. Instead it paid a dependable dividend, typical of mature companies at the time. It had hundreds of offices providing personal loans, vehicle leasing, and life, health, and casualty insurance to small businesses and to the public. It also owned ten small manufacturing companies, acquired cheaply during World War II, which were allowed to operate with little oversight. Once announced, the unwanted takeover offer from Loew's Inc. was a threat to the company's comfortable culture. Control Data, on the other hand, had increased earnings thirty-fold in the previous decade, adjusted for acquisitions, and even more without the adjustment. This difference in energy and ambition suggested that if the companies merged, Control Data would quickly become the superior force. It would be the whippet joining forces with a bear, two different species, and the whippet was considered a "white knight."

Control Data was attracted to this arrangement because Commercial Credit would give it a base of badly needed stability, particularly financing stability. It served business financial needs in many ways and had, in the opinion of Control Data's vice president of finance, "far more resources than are currently needed." Control Data saw itself using these resources both for its internal needs and for its customers, who were increasingly leasing their machines. It also saw an opportunity to reach smaller businesses as customers for its computer products and services. Commercial Credit, on the other hand, saw the Control Data customer base as a natural, captive market for growing its lease business, and it was very much interested in an alternative buyout offer to reward its shareholders. After some dickering, Control Data offered .45 of its shares for each Commercial Credit share, an agreement valued at $750 million. This was far more attractive that the previous, unwanted offer of $430 million from Loew's, but justified somewhat because the Commercial Credit shareholders would no longer receive an annual dividend of $1.80 a share, something they had enjoyed for many years. Loew's declined to make a counter-offer, and on August 15, 1968, the shareholders of Commercial Credit accepted the Control Data terms.

The acquisition was made with "Chinese money" for approximately twenty-five times Commercial Credit's expected earnings, and, once the companies were combined, the remaining company, Control Data, sold at around fifty times the now-consolidated earnings, a more reasonable multiple. Offsetting this cheaper valuation, however, was the fact that at least half of the earnings base was not growing. Nevertheless, the acquisition was considered a significant coup for Control Data. The *Minneapolis Tribune* in a Sunday editorial on August 18, 1968, said few would have dreamed that a new Minnesota company formed hardly more than a decade earlier would itself "become a giant." The company's success had meant "thousands of jobs for Minnesota's economy.... (and) nice stock market profits for thousands of our citizens." Norris and his family held "stock worth more than $60 million." Control Data was called "a modern Horatio Alger success story...(with) new chapters added each year." Helped by the acquisition of Commercial Credit, its book value almost doubled from $22 a share to $40 a share (the acquisition was not "pooled;" it was treated as an "investment" for accounting purposes.) Because Commercial Credit was such a dominant factor in the financial statements of the acquiring company, Control Data changed the end of its fiscal year from June 30 to December 31, the accounting year for Commercial Credit. This change helped maintain the historical integrity of annual comparisons for Commercial Credit, an important consideration since the company was highly dependent on the approval of rating agencies to maintain one of its great corporate strengths, a prime credit rating.

Somewhere around that time, the stock of Control Data traded as high as 174, which, accounting for splits, represented an increase of 783 times its initial offering price barely eleven years earlier. In other words, an opening investment of a thousand dollars would now be worth over three quarters of a million dollars.

But Norris was not about to celebrate. He had been brought low in the previous years, wounded, and the perpetrator of that experience, IBM, was still capable of breaking his run, still looming at every competitive crossroad, envious of his success and wanting to claim it for its own. To create a clear path for his now strengthened organization, he needed to change something fundamental in its competitive position. He had done what he could to build his own operation by developing multiple revenue sources, and then by acquiring a huge financial base. His next big move would be to try to change the industry itself. He would be successful at this, but the stock would never again reach 174. In fact, six years later it would trade below 10.

6

A Sleeper

Well before the local stock market came alive in the late 1950s, bringing many new companies into existence, one organization had already been building a performance record, but it was mediocre at best and gave no hint of the success to come.

The business was founded almost casually in April 1949, when two young men looking for adventure and opportunity decided to start a little operation of their own. One had general management skills and enjoyed mixing with people; the other had technical aptitude, preferring the quiet satisfaction of accomplishment alone at a workbench. They suspected there would be a market for servicing electronic products. They had limited capital and no financial standards, no expectations to be met other than generating a salary and paying bills. In the first month they took in $8 on service calls. Revenues increased over the next ten years, but the company's limited goals meant meager profit.

But once Control Data brought the local stock market to life, making it easier to get financing for expansion, the little company seized the opportunity by issuing debentures convertible into stock. The arrival of a new class of owners and an independent board of directors lead to the realization that the company would now be held to stricter standards of financial performance. That discipline would eventually help shape it into a real business. To the general public, however, Medtronic was just one of many start-up firms competing for investor attention, perhaps another

Space Age name with overblown intentions. It did not receive much interest. But some few investors probably knew that the fellow at the workbench had a hand in saving "blue babies" (see p. 117) and reasoned that there was the possibility of future success.

Earl Bakken was born on January 10, 1924, in Columbia Heights, a suburb tight against the northeastern border of Minneapolis. He was *practically* an only child, he tells us in his autobiography, and he had the run of the house. His father worked in the office of a nearby manufacturing company, and his mother served as secretary of the First Lutheran Church, three blocks from home. His only sibling, Margorie, was born eighteen years later. Such independence led to an early interest in experimentation, and young Earl built a robot five feet tall. He used his erector set for framework and plywood to finish the exterior. The head moved up and down, it had blinking eyes, and it smoked Lucky Strikes. Bakken had constructed lungs from a hot water bottle and made them breathe with an electric motor hid away inside the body. It was not quite Pinocchio, but pretty close. (Almost everyone smoked in those days, including Pinocchio when he fell among bad companions.) Maybe the startling creation was meant to be the brother he never had, or maybe an early phantasm of his life's work, but in a general way it appeared to be a precursor of something special. When fall came and he adapted the robot to Halloween festivities by putting a knife in one hand and having it move up and down, his mother made him dissemble the creation.

Bakken also wired up firecrackers and set them off from an attic window—an experiment that might have led to an altogether different career—but the caper turned out to be only for fun. He also set up a phone system with a friend across the street. He dug tunnels. He made a radio from a crystal set that he kept in his bedroom and played softly at night when his

Young Earl Bakken at his workbench

parents thought he was asleep. His mother encouraged him in his scientific pursuits. She once let him use her hot-plate in an experiment, and she sometimes shopped for old electronic parts he might find useful. She, herself, had superior skills at math and quietly reveled in her son's curious adventures.

Bakken was a regular attendee at Saturday afternoon matinees at the Heights Theater on Central Avenue. One day he saw the movie *Frankenstein*. Most people considered it a horror movie and delighted in the eerie fright, but young Bakken, about eight or nine years of age, was fascinated by the magic of electricity playing out on the screen. The central character, Dr. Frankenstein, brought inanimate body parts to life, andBakken observed that when electricity flows, we're alive, and when it doesn't, we're dead. "I think it is important to record your dreams," he said in later life. "I'm talking about those ideas that come to you in strange times." But maybe he was also thinking of those imaginary moments he had as a young boy in the theater. He frequently mentioned that seeing the Frankenstein movie was a formative event in his life.

Bakken attended high school in Columbia Heights and described himself at the time as "deeply introverted," but he did have some neighborhood buddies. He thought that his eventual work would be in "an anonymous research position deep in the heart of the Minneapolis Honeywell Company, which at that time had a medical division." After graduating from high school, he went to the University of Minnesota for two years; then, like most men at the time, he enlisted in the military. From 1942 to 1946, he served in the Air Force as a radar maintenance instructor, and by 1948 he received his bachelor degree from the University of Minnesota. A superior student, he decided to continue on to graduate school and took three math courses simultaneously. But when he got into advanced thermodynamics in physics, he said "that kinda threw me," and he dropped out of school.

But learning wasn't the only thing on Bakken's mind. Connie Olson, a classmate from high school, found space there as well. She was working as a medical technologist at Northwestern Hospital in south Minneapolis. Girls talk among themselves, of course, and she must have told her co-workers about her boyfriend's proficiency in electrical matters. Naturally, then, when Bakken was waiting for her to finish work, the nurses would ask him if he would do small electrical fixes, as the hospital had no one qualified to work on the delicate electronic devices then new on the market. Bakken, the man who had once built his own radio, a robot, and a telephone system, of course, could fix them. This, it turned out, was the apprenticeship for his life's work, and his subsequent marriage to Connie created the connection to his future business partnership.

His first full-time job was in a corporation. Bakken found it uninteresting. One day at a family birthday party, he was chatting with his brother-in-law, Palmer Hermundslie—their wives were sisters—and learned they both had an interest in creating their own company. They discussed their complementary skills. Bakken believed there was a larger business opportunity in repairing medical electronics. Hermundslie was managing a lumber company in northeast Minneapolis at the time, but he had previously worked at Minneapolis Honeywell as a technical representative for aircraft electronic controls, and at Dayton's department store. During World War II he had been in the Army Air Corps., training pilots. He was a man of varied interests and a smiling sense of humor. Bakken, of course, had a strong technical bent. Hermundslie was thirty years old, Bakken, twenty-five. They drew up a partnership, thinking at first that it was a short-term proposition, and decided to call the new company Medtronic.

Hermundslie's family owned a 600-square-foot garage and woodworking shop, built in the 1930s with lumber removed from a railroad car. It was located at 818 19th Avenue in Northeast Minneapolis, alongside his parents' house in a working-class neighborhood of narrow lots and two-story houses. An oil-burning stove provided winter heat. Hermundslie suggested that Medtronic set up shop there, rent free.

Bakken would contact hospital employees he knew, get an assignment to fix some electronic equipment, and do the work at a bench in the shop. Hermundslie would help in every other way. When the business ran short of funds, Hermundslie "mortgaged nearly everything he had" to help cover expenses. Initially, he took no pay and kept his day job at the lumber yard for living expenses. Bakken had his wife's salary to help with everyday needs.

The new partners accepted jobs wherever they could find them. Sometimes a hospital employee would provide a rough sketch of modifications that would allow an existing piece of equipment to be used in a special way. Back in the garage, Bakken would solder various parts together to create a workable answer. No federal review or approval was required at the time. The little company made a number of those "specials" in the

The garage where Medtronic got its start.

early years and lost money on most of them, Bakken said later, as they invariably underestimated their costs.

Searching for more reliable sources of revenue, the partners in 1950 decided to become sales representatives for Sanborn Co., a medical products firm located in Boston. With this decision, Bakken, the technician who enjoyed experimenting at the workbench, and his partner, the former lumberyard manager, began selling electrocardiographs, physiologic recorders, and other medical diagnostic equipment to doctors and hospitals. More than half of their sales in the early 1950s came from these products. The company grew to half a dozen employees. The makeshift garage was enlarged and a furnace was added; also a toilet that replaced a coffee can. There were no federal clean room inspections at the time. Though growing, Bakken said, they were still "a company in search of a mission." Sometimes, "to keep eating," they simply serviced TV sets. Nevertheless, he recalled, "when we weren't fretting about survival, we realized we were having a wonderful time."

The various product offerings allowed the company to become well acquainted with doctors and nurses throughout the Midwest, including staff members at medical research labs.

Among other things, Bakken worked on early recording and monitoring equipment for anesthesiologists. His activities at the University of Minnesota medical school became so frequent he eventually obtained his own locker in a room with the surgeons, many of whom later became heads of surgery in hospitals around the world. In the mid-1950s, he met Dr. C. Walton Lillehei, the bold pioneer in open heart surgery who wanted someone knowledgeable about electronics standing beside him in the operating room. Lillehei had done the first repairs on a "blue baby," that is, one not receiving enough oxygen due to a congenital heart defect. Restoring such a baby to the full blush of health was a near miracle and was acclaimed as such in the press. But some of the babies left surgery with a heart rate too slow to survive the trauma of surgery, as silk sutures interfered with the heart's normal electrical impulses. A pacemaker would help amplify the function of the heart until it sustained itself, but this was not an easy solution to apply. Pacemakers at the time were made with vacuum tubes and took up a few cubic feet of space. They had to be wheeled about in a cart and plugged into an electrical outlet. A long extension cord was required to wheel a patient out of the operating room, and if the destination was too far away, another cord had to be plugged in farther down to take over when the first cord reached its limit. There were no outlets in elevators or on stairways, or outdoors, or in transport vehicles that brought the patient home. There was also worry that the device might create a pulse too powerful for vulnerable infants.

In October 1957, when the electrical power failed for a period of three hours, the university hospital went dark. There was back-up power in the operating rooms, but not in the patient rooms. All patients were vulnerable, particularly infants relying on electricity to keep a pacemaker running. In a recorded interview, Dr. Lillehei says, "I don't think any died, but they got in trouble all right." Another source says there was a "rumor" of death, and still another says it was a "near catastrophe." Nevertheless, a more reliable source of power was obviously necessary. Dr. Lillehei asked Earl Bakken to adapt the pacemaker of the day to work under more demanding conditions.

Bakken's biomedical effort continued a long line of experiments with electricity and life extending back to Benjamin Franklin's time. One of the more interesting early reports concerned a Danish physicist who in 1775 placed "electrodes on the sides of a hen's head and applied an electric discharge, which made the hen fall dead. When the electrodes were applied over the hen's chest, it staggered onto its feet and walked away."

In 1926, electricity revived a stillborn infant in Australia. One pole was applied to a skin pad and the other, an insulated needle, was inserted into the heart. In the 1930s, an American physiologist, Dr. Albert Hyman, developed an "artificial pacemaker" that was cranked by hand. The name "pacemaker" stuck, but there was negative public reaction at the time about interfering with nature by "reviving the dead," and Hyman did not publish any data on the use of his pacemaker on humans.

Around 1950, a Canadian electrical engineer, John Hopps, developed a large electrical machine using vacuum tubes that provided transcutaneous pacing, using two metal discs placed on the left and right sides of the chest. It was painful to the patient and required sedation, and because it was powered by an AC outlet, it carried the hazard of electrocution. Also, prolonged stimulation produced skin burns.

In 1957, Dr. Lillehei and his associates at the University of Minnesota developed a stainless steel wire in a sleeve that went right to the muscular substance of the heart and was able to work with much lower voltage. Once the heart had resumed normal conduction, the wire was removed.

It was later that same year that Lillehei asked Bakken to

develop an improved pacemaker for "blue babies." Utilizing transistors rather than vacuum tubes and circuitry designed for an electronic metronome, Bakken built a battery-powered and therefore portable external pacemaker, four inches square and an inch and a half thick—small enough to be hung around a patient's neck or strapped to the chest. It had external controls permitting appropriate adjustment to the rhythm of the heart. The unit was no longer vulnerable to a sudden loss of power and the wearer was no longer tethered by the length of a cord. It weighed approximately half a pound. No patent was filed on the invention. (Medtronic now holds more than 22,000 patents worldwide.) Earl was surprised to see the device used almost immediately without additional testing, but Lillehei explained that when a patient was in crisis there was no reason to wait. The unit was originally used for the temporary pacing of children.

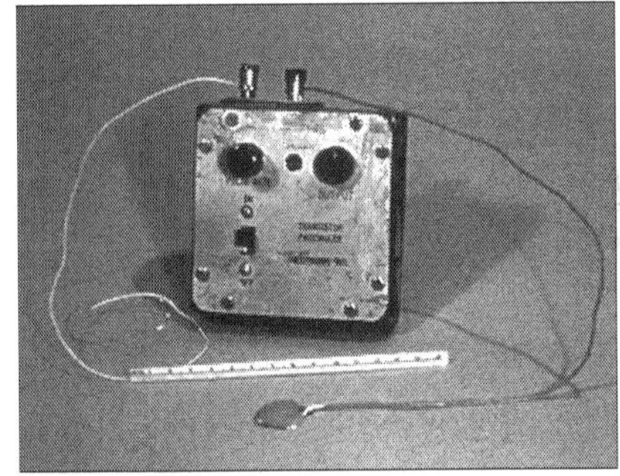

The first Medtronic Pacemaker

With improvements, the device was made suitable for permanent pacing through electrodes that could be sutured directly to the heart. In March 1959, at St. Joseph's Hospital in St. Paul, Dr. Samuel Hunter installed the first one in a seventy-two-year-old man who recovered and lived an active life for six or seven years. (Accounts differ regarding the date of his death.) This life-saving development was extraordinary but not ideal, as wires through the skin might become dislodged during sleep or break from strenuous physical activity; also, there was the possibility of infection.

At about the same time, an *implantable* pacemaker was developed in Sweden, using a short-lived battery that could be recharged each week through the skin. The device was implanted in a forty-three-year-old patient who ultimately received almost thirty replacement devices and died at the age of eighty-six from cancer.

Today, a Lillehei quote concerning medical advances is memorialized on the walls of the Lillehei Heart Institute at the University of Minnesota: *What mankind can dream, research and technology can achieve.* Earl Bakken was his technologist.

Medtronic had six employees and generated $40,000 in revenue in 1957, its eighth year in business. The company had worked on various medical innovations but didn't have a thriving business. The two partners decided to become incorporated, perhaps to limit liability, and converted their ownership into stock, but not for the public market. Bakken, then age thirty-three, was named president, and Hermundslie, thirty-eight, executive vice president. Both served as directors. According to one source, Thomas Holloran, a Minneapolis attorney, was named as an outside director at the time, but another source says he became a director in 1960. Holloran today confirms the later date. The other officers were Leslie Kotval, treasurer and comptroller, and James Anderson, vice president of sales. The company was staffed to be successful and its work was promising, but the financial results were marginal.

Needing capital, in December 1959 Medtronic prepared an offering of 5 percent subordinated debentures, convertible into common stock at $1.50 per share prior to July 1, 1960, and higher prices thereafter. If investors converted in the earlier period, they would own approximately one-third of the company, and Bakken and Hermundslie would own equal shares of the remaining two-thirds. The prospectus was fourteen pages long including the subscription agreement; its goal was to raise $215,000. Company financial statements showed a current

Palmer Hermundslie

cash position of $320 and various notes and contracts payable of $39,518. Over the years the company had taken on sales representation for various other companies, and the prospectus recited fourteen different product offerings currently available, plus thirteen in field test, seven in development, and six proposed for future development; the array of offerings suggested that Medtronic was attempting to meet every customer need. But sales in the most recent fiscal year had been only $142,000, and net profit $1,500. The most significant revenue item was the 5800 pacemaker developed at the request of Dr. Lillehei.

The company engaged a brokerage firm to do the underwriting, but before any debentures could be sold, the brokerage went out of business. Like the original employees of Control Data two years earlier, Palmer Hermundslie then tried to sell the issue through personal effort. He called on businessmen along Central Avenue in northeast Minneapolis and succeeded in getting orders at $1,000 a debenture, but not enough to complete the deal. Other brokerage firms were invited to help with the offering. Tony Gould, a stockbroker at Continental Securities, remembers his colleagues traveling to the Mayo Clinic in Rochester, Minnesota, trying to interest doctors in the new issue, a very logical appeal to individuals who would understand the company's possibilities, but the brokers returned with no orders. By fiscal year end, April 30, 1960, however, through various efforts, the offering was fully subscribed.

Since it was advantageous to convert within the first six months at the $1.50 per share rate, some investors did convert their debentures into common stock, and a quotation under the name Medtronic first appeared in the stock tables of the *Minneapolis Tribune* on May 15, 1960, at 2.4 bid, 3.4 ask.

With the market price exceeding the debenture conversion price, the entire offering was soon converted. Owners could now sell some of their holdings if they wished, and, alternately, members of the public could buy if they wished. A year later there were more than eight hundred shareholders. In this indirect way, the conversion brought the company to the public market and created a measurable value for the business. Once all the debentures were converted, there would be 413,190 shares outstanding, meaning the total worth of the company, based on the initial market quotes, was $1 to $1.4 million. Today, Holloran says Medtronic could not have succeeded in its offering without the unusually receptive market created by the early success of Control Data. It took a leap of faith to believe the company could eventually become a financial success. The first annual report of the now-public company said that the past year had been "tremendously interesting," and further, that an enlarged sales organization and a planned new facility led to anticipation of "an eventful year." Sales were $180,000 and the company lost $16,000.

Meanwhile, in Buffalo, New York, an electrical engineer named Wilson Greatbatch was working with Dr. William Chardack on a pacemaker that could be surgically implanted. For two years Greatbatch had done experimental work, and in early 1960, Dr. Chardack completed a successful implant at the local veterans' hospital. Medtronic was familiar with this work since the developer had expressed interest in a Medtronic electrode—one that could be sewn directly onto the heart. In October of that year, Palmer Hermundslie flew his private plane to Buffalo in order to negotiate a license agreement for the new product and offered to pay a 10 percent royalty. A week later, he and Tom Holloran, the attorney, again flew to Buffalo to sign the agreement. Thus, in 1960, Medtronic was able to offer implantable pacemakers to go along with its external pacemakers. The internal units lasted a little over a year before requiring replacement and cost around $375, about the same as the company's external pacemaker. Finally, after a dozen years, Medtronic was "beginning to discover its purpose," said Earl. It was becoming a real company in the emerging field of medical technology. The stock ended the year at 2.5 bid, 3.2 ask, almost unchanged from its initial quote earlier in the year.

Thomas Holloran

In late February of the following year, something sparked a burst of excitement in the stock, and it hit a bid of 16.4. Two weeks later it fell back to 9, and six weeks after that, it sprang back to 15.4. It was as though the stock itself had received a couple jolts of electricity. Did local investors suddenly recognize an impressive, overlooked opportunity right in their own community? Possibly. But more likely they were reacting to a magazine article that gave the little company national exposure.

On March 4, 1961, the *Saturday Evening Post* published a major article referring to pacemakers as a "transistorized spark of life," and mentioned that close to a hundred individuals, including some in Europe, South Africa, and Australia, now had new life because of these devices. The article featured a large picture of Dr. Lillehei checking a patient wearing a Medtronic device.

In April 1961, Medtronic moved into 15,000 square feet of new space at 3055 Old Highway 8 in St. Anthony, just beyond northeast Minneapolis. It now had more than fifty employees. Picker International, a much larger company, had become the company's overseas distributor. This meant that Medtronic was represented in more than seventy offices around the world, complementing the fifteen independent companies providing representation in the U.S. and Canada and the three direct sales personnel who covered the Midwest. "Medical electronics is still an infant industry," Bakken said in a public statement. "Its potential market has barely been scratched."

Although the opportunity was large, cash was scarce. The money raised in the initial debenture offering had been spent, and an additional $133,000 had been borrowed from a neighborhood bank. Large salaries were not the problem, as Bakken was making $9,600 a year, and Hermundslie, just $8,400.

In October, *The Upper Midwest Investor* published a lengthy feature article on Medtronic, reciting its history, its medical accomplishments, and its continuing opportunities. It also quoted Bakken saying that "a business must also show a fair profit," something the company had trouble doing. By year end, the stock had drifted back to 3.6 bid, 4.4 ask, up approximately a dollar from its close a year earlier.

Sales in the fiscal year ending April 30, 1962, increased to $518,000, but because of ambitious spending the company reported a loss of $144,000, up from losses of $53,000 and $16,000 in the preceding two years. The local bank, uncomfortable with its loan, had shifted responsibility to its larger downtown affiliate, Northwestern National Bank. Medtronic then borrowed $100,000 from the downtown bank on a promissory note secured by accounts receivable and inventory, and issued new debentures in the amount of $200,000 to Community Investment Enterprises, a local venture capital firm. In return for the financing, the two organizations insisted on board representation. William Dietrich and Gerald Simonson from Community Investment Enterprises joined the board that year, and Donald Shultheis from Northwestern National Bank joined a year later. Earl Bakken said later that the advice from these new board members concerning financial controls and oversight was more valuable to the company than the money received.

In an effort to restore financial balance, half the company employees were laid off and both Bakken and Hermundslie went for a year without pay. Product lines that didn't fit with the core strategy were phased out. An effort was made to sublease space in the new building, but there were no takers since it was 1962 and the business environment had cooled considerably. The bank suggested that the company put itself up for sale, and the rest of the board, including both Bakken and Hermundslie, agreed to consider the idea. A search of possibilities produced just one interested buyer, Mallory Battery Company, a supplier to the company. Other candidates looked, but showed no interest. Mallory then made an offer to buy the company at $3 a share, subject to further review. "We were desperate enough, despite our growth and advances, to give the offer serious thought, figuring that if we did sell out, we'd at least get some money out of the crazy little business to which we'd devoted the last dozen years," Bakken reminisced later.

Mallory Battery Co. commissioned the consulting firm Arthur D. Little to estimate the potential market for pacemakers. After completing their study, Little concluded there was a worldwide opportunity for about ten thousand units. Apparently considering this insufficient, Mallory withdrew its offer. The struggling company was then left to make its way alone, trying to convince medical doctors, one by one, to try the new technology. Nevertheless, the officers were pleased to learn about the number of selling opportunities still ahead. Their own sales had been increasing and they knew there was growing interest, but until then they had no way of quantifying it. The results of the study confirmed their impressions.

Realizing they now had to manage their own destiny, the company directors told Bakken he had to stop tinkering with electronics and instead embrace the role of chief executive, or find someone else to run the company. The choice gave him pause. He said it "necessitated, of course, constant speaking and making difficult decisions," but he concluded that he really "wanted to run the company."

Admitting a certain discomfort among strangers, Bakken began to work on his skills. He practiced the techniques of public speaking. He replaced his plaid shirts and work pants with suits and matching ties. He read everything he could on management. Friends introduced him to golf. He hit the ball quite well, one observer said, and had the fluid movement of an athlete, but when asked to play a second time, he replied, "No thanks, I've already done that."

And he took lessons in ballroom dancing. Roughly six feet tall, he moved gracefully with the sway of the music, easily picking up the count, and discovered that this was an activity he enjoyed. His continuing improvement carried him to the highest (gold) level of dancing proficiency. When his first marriage failed after thirty years, Bakken and his dance instructor got married. The self-described introvert had transformed himself into a ballroom personality, and became at the same time an accomplished executive. "If I changed at all, it's because of the necessity to run the company. It isn't what I planned, that's for sure," he observed later.

As early as 1960, Bakken had been working on a corporate mission statement. Now that new directors were more actively involved, he gave them an outline for review and approval, and it was formally adopted in 1962. The completed document began with a near supernatural aspiration to "alleviate pain, restore health, and extend life," ideals that reflected the founders' personal attitudes, almost a spiritual calling. This was followed with business specifics. The company would direct its growth to "areas of biomedical engineering where we display maximum strength and ability" and at the same time "avoid participation in areas where we cannot make unique and worthy contributions." Also recognized was the need for "a fair profit . . . to meet our obligations . . . and reach our goals." The overall statement, an exposition so precise and clear it took only half a page, identified the company's direction for itself and also helped convince all interested parties that Medtronic was serious about creating a successful business. Phasing out various products in 1962 reflected the new corporate discipline. But investors at the time saw only a company losing money. By year end 1962, the stock dropped to 1.6 bid, down 90 percent from its high the previous year. That fall, Bakken had gone on a medical electronics trade mission in Europe to help open up foreign markets. Palmer joked that when Bakken left the country, the company started to make money.

In the fiscal year ending April 30, 1963, sales reached $985,000, almost double a year earlier, and the company reported a profit of $73,000, or 16 cents a share. "We anticipate that the total market for prosthetic equipment will more than double each year for the foreseeable future," Bakken said confidently, and added an embellishment: "The future of Medtronic looks limitless." But investors either were not paying attention, or they were skeptical. At mid-year 1963, the stock traded at 2.5 bid, 3.3 offer, meaning it was about 20 times earnings—very cheap for a company about to double its operating results on a continuing basis, and far short of Control Data's 100 times earnings when it was undergoing similar growth. But the Medtronic product was new and the company's operating history had been troubling. Gradually, however, a new realization set in. By year end, the stock more than doubled to 5.5 bid, 6.2 offer.

In fiscal year 1964, sales hit $1.6 million, and the company earned $198,000, or $.26 a share, up from $.16 a share in the previous year. In the following year the stock traded as high as $11. Thus, after the severe financial squeeze in 1962 and an unsuccessful effort to sell the business, the company was now performing well. As Bakken had earlier suggested, sales continued to advance and reached $9.9 million by fiscal year 1968. Earnings per share, after a 2 for 1 split, had increased tenfold from four years earlier. Such expanding results called for increased capacity, and the company began construction of a new manufacturing facility in Holland.

In 1968, the stock hit $55, which meant it sold at approximately 70 times earnings. Adjusted for the split, the price had increased 73 times over the initial debenture conversion price just eight years earlier. Bakken was performing as a corporate leader, and the results showed it. Tom Holloran, an advisor to Bakken in many ways, said later, "This quiet, thoughtful, brilliant man was very captivating. He has been the great mentor in my life... he persuaded me to leave the law firm and take a pay cut," adding as further evidence, "at the time... there were no bonuses paid." Holloran later succeeded Bakken as president of the company.

Further testimony to Bakken's leadership came from a mid-level administrator who had joined Medtronic from Picker International in 1967. Young Manny Villafana said that Earl Bakken was a very good teacher. In time, as his career progressed, the eager student went on to create seven medical companies, including the huge, thriving, present day success St. Jude Medical. Bakken's executive influence was spreading quietly throughout the Twin Cities business community.

But while Bakken was gaining executive stature, Palmer Hermundslie was struggling. Because of diabetes, his eyesight was beginning to fail, and then his kidneys. The man who had given so freely on so many occasions, and whose work had helped improve the lives of so many, was unable to reverse his own condition. Still, he continued to work in whatever way he could. But in December 1969, the illness forced him to retire. No doubt friends and neighbors who had bought the initial debentures at Hermundslie's suggestion thanked him for their good fortune, gave him a good handshake, and reminded him that his work had been invaluable. A few months after retiring, Palmer Hermundslie died at the age of fifty-one.

Looking back, it could be said that the simple financing Medtronic secured in the local over-the-counter market in December 1959 created the necessary discipline for success, and ended up fueling an outpouring of medical innovation that improved lives, created employment for many thousands, and brought accompanying investment rewards. It had been a long and difficult incubation, but perseverance and a standard of excellence produced good health and good fortune for many. Today, in the United States alone, 230,000 pacemakers are implanted each year. Worldwide, the number exceeds 600,000.

7

A Meteor

Then came Quarterback Sports Federation, Inc., a stock financed by a local brokerage firm. The name of the new company sounded like a group of quarterbacks collaborating in some way; maybe a Monday morning gathering of would-be athletes kibitzing the weekend games. But what's a federation, anyway? A league? And what business are they in? It might involve more than football, since *sports* is an all inclusive word. Grandiose rhetoric seemed to be making a comeback in company names, but when the public learned that Quarterback Sports Federation was in the restaurant franchising business—that is, serving food—everyone felt they understood.

Stockbroker Tony Gould remembers that the company was originally set up to sell deodorants and cosmetics but decided to switch to hamburgers. Such a move meant the company would be an early entry in the same business McDonalds was succeeding at with many hundreds of outlets, and this was generally known since Ronald McDonald was already appearing on television in 1966. A&W, another popular chain, might have been considered a competitor, but it championed root beer. Thus, the new company represented opportunity. Investors had the understanding that Vikings football players, including quarterback Fran Tarkenton and teammate Bill Brown, were early investors in the stock. This increased public interest and gave some credence, however slight, to the company's name. After the offering there were a little over 100,000 shares outstanding—a quantity large enough for orderly trading if the public showed moderate interest, but also small enough to be manipulated by a cabal of market players should they decide to do so.

A market quote on Quarterback first appeared at 1.4 bid, 2 ask, in the Minneapolis newspaper on February 8, 1966. Two weeks later it was at 6.4 bid, but by August it had dropped to .7 bid. The company reported revenues of only $34,000 in the fiscal year ending June 30 and lost 25 cents a share. A person who bought at the $2 level probably held his breath and wondered, *Is this going to be another balloon gone soft?* But by year end, the stock recovered to 2.2 bid. *That's better,* he likely thought. *I'm ahead now. Might as well let it play out.*

One of the company's first restaurants was on Oak Street near Memorial Stadium on the University of Minnesota campus. The roof of the building curved to a point at the peak, suggesting half a football. *Catchy,* the investor thought, *and a good location too. Students should love it.* Word began to circulate that the company was negotiating franchises. By the following May, the stock had reached 8 bid, a new high. *If I sold now, I would be making four times my money in a little over a year,* the investor calculated. That evening he said to his wife, *I own this little stock called Quarterback. Some of the Vikings are involved. So far it's up four times. What do you think?*

Sell it, she said. *Be grateful.*

But they're offering franchises for restaurants. Everyone has to eat. Who knows how many they might open? It could go a lot higher. It could be another McDonalds.

Well, maybe you're right. she said.

He decided to hold on.

That summer the stock hit 15. He called his broker. *They opened another restaurant in Savage, out along the Minnesota River,* the broker said. *Have you seen it?*

No, but I've seen the other one. It looks good. And I hear they're opening one on Cedar Avenue in south Minneapolis, and another in

Northfield. They offer three hamburgers: the Quarterback, a quarter pounder; the Half Back, a half pounder; and the Full Back, referring, of course, to Bill "Boom Boom" Brown. Neat, huh?

Can you imagine when they start opening these all over the country? It could really be big, the broker said. *I'm a believer. But the stock might be getting a little ahead of itself. You should consider selling some. Once it backs off, you can buy in again.*

The investor wasn't sure what to do, so he held on. *My cost is 2,* he told himself, *and the rest is gravy.*

In October, the stock hit 24. The company had finished its second year and reported revenues of $385,000. They had now set up a territorial franchise, Viking Quarterback, which gave the right to open hundreds of restaurants in the Midwest. Those seeking such a franchise paid advance money, which dropped right to the bottom line, and the company reported earnings of 50 cents a share.

Home for supper that evening, the investor mentioned the accelerating price to his wife and then asked, *What do you think we should we do?*

Sell it, she said. *Consider yourself lucky.*

You said I should sell at 8 last May. Now it's three times higher.

Well, she sighed, *suit yourself, I'm no expert.*

The stock finished the year at 41. Now the company was discussing two more territories, Metropolitan Quarterback and Quarterback East, both on the East Coast. Star NFL quarterback Johnny Unitas was reportedly involved. A separate subsidiary was set up to provide consulting services for the equipment used in company restaurants and also other restaurants. *Holy cow, that in itself could be huge,* the investor thought. At brokerage desks in the Minneapolis and St. Paul downtowns, orders to buy the stock were coming in from around the country.

What should we do? he asked his wife. *It's now at 41.*

I don't know, she said.

Photo courtesy of the Northfield Historical Society

A Quarterback Club restaurant in Northfield, Minnesota

I thought about selling, but then we'd just have to pay taxes.

He decided to hold on.

In January, the stock hit 55. He wasn't sure what to do, so he called a friend who kept charts of stock prices and therefore always seemed to speak intelligently about investing.

I've never seen a better looking chart, the friend said. S*traight up for over a year. There's no overhead resistance. I'd hang on.*

Let your profits run, is that what you mean?

That's right. Let your profits run, and cut your losses. The best advice I can give.

He liked that and decided to hang on.

Two months later, the stock dropped to 33.

It's down 40 percent, he said to his wife at supper. *That's not good.*

Sell it, she said, *33 is still a good price. You only paid 2.*

But I feel like I'm losing 40 percent. If it runs back up, I'd regret it. Somebody thought it was worth 55 in January.

He went back to the friend with the charts.

I see no underlying support, but I suspect it's going to form a base at this level, the friend said, while examining the pattern of price and volume. Then he looked up to give a summary opinion. O*nce it moves off the base, you will be able to tell its future direction.*

I hope you're right, the investor said.

A month later the stock moved up to 42. He was pleased he had held onto his position and wondered what to do now, but he knew for sure he wanted to thank his friend.

The friend looked at an updated chart. *It rallied just like I suspected. Let your profits run. It's moved off the base now.*

He liked that idea and decided to hang on. *It's free money anyway,* he told himself, and grinned in satisfaction about his good fortune. But still he worried, because it had become a sizable dollar amount. One week later, the stock split 2 for 1. It dropped to 22 bid, but he had twice as many shares. He called his broker.

More shares mean more liquidity, the broker said. *When that happens, shareholders like to sell half their holdings and pocket their gains. You should consider that.*

What do I care about liquidity? he thought, and decided to hold.

The stock rose quickly to 36. *Not bad,* he said to himself. *I'm glad I held.*

But two months later it was at 22. He decided to call his broker. *Did it split again?* he asked.

Nope, it just hit a downdraft. I told you to sell half.

He kicked his leg as though dismissing an unwanted thought, and quickly ended the conversation.

Six weeks later the stock had streaked to 54. *Amazing,* he thought, and called his broker looking for guidance.

Who knows with this stock, the broker said. *But I hear good things about the company. They reported their results. Did $2.5 million in revenue and earned $1.71 a share. Wow!*

Good. I'm glad I held, he said. *Someone told me the company is opening two more territorial franchises: Southwest Quarterback, and Quarterback Clubs West, on the Pacific Coast. I hear Roman Gabriel is involved. He's the Los Angeles quarterback, you know. I'm hanging on.*

And the company set up another subsidiary called Dart Investments to engage in restaurant real estate development. In February 1968, some Dart shares were offered to the public at $3 a share; but Quarterback continued to hold a sizable ownership position. By year end, the Dart stock was selling at 41. Quarterback stock moved apace, going from 54 to 105 in a steep ascent over the last three months of 1968 based largely on its holdings in Dart. *I bought the stock less than three years earlier at 2 a share,* he told himself, *and now hold twice as many*, so he did the arithmetic. *From 2 to 210, this is real money.* He grinned and bought himself a candy bar to celebrate.

That night he and his wife discussed their sudden fortune.

Let's sell and buy a place up north, she said. *We could leave this rat race and take up fishing.*

Fishing only goes so far, he said. *We should consider this our retirement nest egg. I know a person who lives off his dividends. That could be us someday.*

But he wasn't quite as sure as he sounded. He called his broker to talk further. *What causes this stock to go up so much?* he asked.

They plan to create subsidiaries of the territorial franchises, keep a majority ownership, and sell stock in the subsidiaries in a public offering, the broker said. *Just like Dart, the real estate subsidiary. Then the parent company, Quarterback, will hold stock in all these subsidiaries worth lots of money. That's what's driving it. And it's a formula they can keep on replicating. More subsidiaries trading at high prices mean more value to the parent. The public loves this stuff.*

Interesting, he said, *I like it too.*

The stock started to slide. *A downdraft,* he told himself. *Just like before. I'll wait this out.*

But he felt uneasy as he watched the descending quotes. Into the 70s in mid January. Down to the 50s in February.

And then it stabilized...and then it rallied. It hit 62 in late March. *Like I suspected,* he said to himself. *I've seen this before.*

A day later it dropped to 31. He called his broker. *Did it split?* he asked.

Yes, 2 for 1.

That's a good thing, isn't it?

It increases liquidity, the broker said. *More liquidity in a declining stock. I'm not sure how to assess that, but it doesn't sound good.*

He hung up, unsure what to do. *Another split,* he said to himself. *That sounds like a good thing. I've now got four times my initial purchase of shares.* And he thought of the great leverage he would have once the stock recovered to previous levels.

A month later the stock was at 20. He felt sick about it and didn't want to bring the subject up with his wife.

But the stock rallied again. In May, it hit 29, up almost 50 percent. He went to the friend who kept the charts.

I don't like it, the friend said, studying the price pattern, which showed a steep drop from a rounded top and then a recovery. *That looks like the right shoulder on a head-and-shoulders formation. Not good.*

He was bewildered now, unsure what to do, so he asked his wife. *What do you think?*

I have no idea, she said. *Maybe it will fund our retirement.* He thought he heard a trace of sarcasm in the remark, but the voice came with a smile, so he wasn't sure.

He decided to hold. *After all, we only paid 2 for it,* he told himself.

It started dropping again ... to 15 in mid-June and then to 11 two weeks later.

In the first week of July, it sparked up again, back to 15.

But by August, it fell in half, now at 7. He hated checking the price. When doing so, he put his hands to his head so he could think better.

A week later, the stock dropped by half again, now 3.5. *If*

I sell now, it's not worth much money. And it might go back up, he reasoned.

Which it did. By August 30, it had doubled. *Whew,* he felt relief.

But it didn't hold at that price. There was liquidity, all right. On the down side. *Not enough buyers to drink that liquidity,* he thought.

The stock dropped to 2 in November. *I'm even,* he said. *No point doing anything now.* But then he remembered he had four times as many shares.

As the year end approached, the stock dropped to .7 bid.

But three weeks into January of the following year, it was 2.4 bid. *What's going on,* he wondered, and decided to call his broker.

They do that sometimes, the broker said, *bounce like a dead cat after the first of the year. It's a recovery after the previous year's tax selling.*

I hadn't thought of that, he said. *Thanks.*

On April 19, 1970, Quarterback stock was quoted at .3 bid, and that was its final day.

There were no more quotes in the days that followed, since no one was willing to bid for the stock. The value of Quarterback Sports Federation was now officially zero. *It's like getting tackled behind your own goal line,* he said to no one in particular, and he kicked a stone out of his path.

But when he told his wife that evening, he chuckled a hollow laugh. *Can you believe it? There is no market for the stock;* the strained humor was a weak attempt to cover his emptiness. *Unbelievable,* he muttered, as much to himself as to his wife. She looked at him to acknowledge his words, but could only shake her head slowly side to side, to convey her own lack of understanding. Together they ate supper. *It didn't work,* he said later, trying to put the subject to rest. His wife smiled; she was disillusioned as well, but she thought it slightly funny, and she looked into his eyes with bemused affection.

That helps, he said

What?

Your kind look.

Earnings were not properly reported, he learned later. Payments for the right to a franchise were not the same as opening a restaurant. The company had puffed up earnings from the receipt of early payments, then built up expenses and quickly run out of money. No doubt, some brokers helped promote the extraordinary run, but many traders couldn't believe that it would last. John Simma, a stockbroker at the time, remembered the day Quarterback stock traded at 104 (bid) and 110 (offer). His firm's head trader came out to the sales floor and said he was short 4,000 shares of Quarterback, and if he didn't cover his position (that is, buy an offsetting amount) right away, the firm could go under. Jim Donovan, a top salesman at the time, was able to get his largest customer to sell enough stock to avert the crisis. Most likely no one realized it at the time, but Donovan had done his client a great favor.

The sensational run and collapse of Quarterback lasted just four years. Few additional restaurants were built, but during that manic performance the stock went from 2 (ask) to the equivalent of 210 (bid) at its peak in December 1968, split again as it was declining and went to zero in the spring of 1970. Its brief run typified the second phase of the Twin Cities local stock market bubble in the late 1960s, but it was not a fable; it was a real company, trading in real dollars.

At least I didn't buy at the top, the investor told himself.

It was a head-and-shoulders formation, all right, but very steep, the chartist said as he reviewed the final pattern.

You can't beat taking a profit, the broker said and dialed up another customer to tell him the same.

Sometimes your first instincts are the best, said his wife, as though she too had discovered an abiding principle. Then she rushed to pull a soufflé out of the oven and placed it on the counter hoping to admire it before it collapsed.

In December 1971, the Quarterback real estate subsidiary entered into bankruptcy. There was some follow-up legal action, but it didn't make headlines. Nobody could prove that the stock price of the parent company had been manipulated, but some suspected it.

New listings, of course, do best when people feel they can make easy money. In the late 1960s a hopeful, bullish mood returned to the Twin Cities over-the-counter market. With it came new listings. There were seventy-nine in 1968 alone, a five-fold increase from the average at mid-decade. Inevitably, the enthusiasm brought some excesses, but Control Data had shown that big money could be made, and investors wanted to believe. A wise Wall Street executive once said, "How does something get away from you? I'll tell you how. Because it's working."

A few quality companies had gone public a few years earlier: Deluxe Check at $21.50 in December 1965; Dayton Corporation (now Target) at $34 in October 1967; and once 1968 got underway, H. B. Fuller and Arctic Enterprises, names that endure even today. At the time Medtronic was also in the midst of its gallop to 50. But as the momentum built, many companies came public that didn't last and at least one that got flushed—stocks such as Big Steer Restaurants; Donnybrooke Racetrack; a karate instruction school named Chuan-Lo Enterprises; and a manufacturer of chemical toilets called Aquazyme Industries

Mike Gruidl, then an investor in various local stocks, remembers owning a company with vast ambitions named World Wide Petroleum. According to reports, it was doing business on University Avenue, southeast of the University of

Minnesota campus. He went over to take a look and had trouble finding the address, but learned with some disbelief that it was the gas station he had been ignoring as he drove around squinting to read the numbers on buildings. He stopped and asked the station attendant about the company and was told, yes, it was located there, in bay three. He looked inside and found six or eight cases of oil piled on the floor in bay three; not just ordinary oil, of course, but oil intended for sale to racetracks around the world. That, he learned, was the company's place of business. World Wide Petroleum did not go on to success.

In addition to the speculators, the second wave of enthusiasm also brought observers who watched with envy from the sidelines. Commercial West published a listing of stocks with extraordinary price increases in the twelve-month period ending December 1967 and called it "The Return of the Local Market."

	Dec. 27, 1966	Dec. 28, 1967
Reuters	1¼	15½
Analysts International	3⅜	20
First Midwest	4¼	15½
Data Products	2⅝	21⅛
Quarterback	2½	42
National Connector	3⅛	22¾
Sterner Lighting	⅞	5⅛
Electro-Craft	1⅛	6¼
Food Corp.	3⅜	10½

Those were some handsome returns, indeed. Considered in hindsight, the author of the article could only say, "wouldn't it have been wonderful" to own these stocks. But the excitement didn't end in December 1967, as the listing suggests. After splitting, Quarterback stock would go five times higher in the following year. Control Data had its spectacular recovery during the same period, trading at 165 or higher in both 1967 and 1968, but it wasn't on the list because by then it was trading on the New York Stock Exchange.

Not everyone, of course, stood on the sidelines watching. Jim Fuller, by then working as a business reporter for the *Minneapolis Tribune*, remembers doing a package of articles featuring individuals who were buying and selling local stocks. The details of one in particular remain in his memory. Fuller interviewed his subject on the steps outside City Hall—a bachelor in his mid-forties, doing menial work and living a quiet, frugal life. The man subscribed to the *Wall Street Journal* and other financial publications and took a great interest in the market. There did not appear to be any other significant activity in his life. He regularly visited the various brokerage houses located on Sixth and Seventh streets not far from City Hall. Starting with a small amount of money and investing primarily in local over-the-counter stocks, he had a reported net worth at the time of $2 to $3 million. Though modest about his success, he was certainly proud enough to comment on it, but for him it was like winning a game of Monopoly. He had no particular desire to otherwise enjoy his fortune, then large enough to allow him to retire to a life of ease.

The rocketing success of the local market soon gave rise to an investor technique called "blind pools." These were stock offerings asking for an investor's money without explaining where it would be invested. Company officers would be trusted to find a good business opportunity, similar to the expectation investors placed in a mutual fund or closed-end investment company, the difference being that only one opportunity would be chosen and the company would not be subject to the special rules governing the other more regulated investment vehicles. Diversification was not an objective; carefully choosing the right company was. Once the managers of the blind pool identified an operating company that could benefit from the cash they had to offer, they arranged a merger, issuing stock for the acquired company. The operating company then had the use of investor funds and was expected to achieve a higher level of success. The owners would have the benefit of a publicly traded stock without the distraction and expense of filing an offering. And the blind pool investors would end up with significant ownership in an ongoing business without having to go through the time and uncertainty of bringing a start-up company to maturity. Such an arrangement had conceptual appeal to those who might want to ride the coat-tails of managers they considered knowledgeable about local opportunities.

George Kline believes that he and P. R. Peterson, both involved in the local market at the time, started the first blind pool, although it began with a public company called Gopher Grinders. The company didn't have anything to do with grinding gophers; it made and sold a device for slicing bread. The machine didn't work very well, but the company had cash and public shareholders, meaning it had some advantages over a new-issue start-up. Once they gained control of Gopher Grinders, the new managers changed the name to Automated Manufacturing Products and then raised additional money through a "Regulation A" offering, taking advantage of an exemption from full Securities and Exchange Commission requirements designed to foster growth in small businesses. The offering would be under $300,000, approved by the state securities commissioner, and not subject to intrastate selling limits.

Kinnard Investments, one of the leading local securities firms at the time, decided that if it went public, it could put the additional money to good use, so it agreed to combine with Automated Manufacturing Products; the shareholders of AMP became owners of Kinnard Investments. Kline points out that blind pools gained popularity after that and helped finance a number of companies, perhaps the most noteworthy being Techne Corp., a biotech firm now operating out of northeast Minneapolis with over $300 million in annual sales, 750 employees, and many well-rewarded shareholders.

As always during a period of rampant speculation, sources of reliable information are invaluable. Recognizing this, an enterprising publisher brought out a weekly newsletter, about six pages long, called *Market Maker News*. Some brokers called it a "tout" sheet, and tried to cultivate the editor in an effort to get exposure for the latest information on their favorite stock, or perhaps to get exposure for *their interpretation* of the latest information on their favorite stock. Once the news was disseminated to the larger brokerage community, the stock would invariably move up, creating a trading opportunity for those who had anticipated the move. Brokers would discuss the success or failure of such maneuvers and who was behind them over drinks after work at the Court Bar on 7th Street, off Second Avenue.

While working for the *Milwaukee Journal* covering business matters, columnist Dave Beal captured the feel of the Twin Cities business environment in a series of articles intended to explain the differences between the two areas. Milwaukee at the time was dependent on foundries, metal bending plants, and old line breweries; its local market did few new issues, investment funds were meager, and technological innovation was rare. In the Twin Cities, on the other hand, he reported that numerous new firms were being launched and financed by investors "in a steamy, sometimes seamy" market environment.

Beal quoted James Bergtold, then heading a small investment firm, who described a frantic scene when investors called to place orders for hot new issues. "I'd have four people on hold at once. It was absolutely crazy. One stock came out at $10 a share and went to $40 in the same day. It was too high at $10."

"There were a lot of bucket shops then, a lot of questionable practices, a lot of companies going public that shouldn't have," said George Bonniwell, who later became president of the brokerage firm, Craig-Hallum Inc.

Donald Soukup agreed, but ended with a mild counterpoint. "At least half a dozen investment firms went out of business because of shady practices. The public got tired of being ripped off. There were a lot of bad things that happened, but there was a lot of good that came out of that time." At the time, Soukup was president of a Twin Cities venture firm.

One of Beal's articles quoted Bill Drake, one of the founders of Control Data, Midwest Technical Development Corp., and later Data Card, who drew attention to the benefits. "It was speculative fever, by some standard, and that was an excess. No doubt about it. However, if you analyzed what happened there, assessed the jobs created, the tax base created and so on, and equated that with whatever the negatives were, you would find it was a huge plus for society."

There was evidence to support Drake's contention, as Pentair, National Computer Systems, Modern Controls (MOCON), Advance Circuits, Circuit Science, First Midwest, and Technalysis all came out of the second local stock market run-up and went on to become successful businesses. Pentair, which started as a weather balloon company, today has more than 15,000 employees and $3.5 billion in revenue, primarily from the world-wide sale of water-processing products and solutions. (Note: an acquisition in 2012 doubled sales to $7 billion in the following year, much of it coming from international markets.)

Underlying this investment fervor was another matter that demanded attention, not only in the Twin Cities but nationally. Because trading was unusually heavy, it strained back-office record-keeping almost to the breaking point, even for the big national firms. New York Stock Exchange volume that averaged 4.4 million shares a day in 1964 almost tripled to an average of 13 million shares in 1968. (In comparison, the number of shares trading on the New York Stock Exchange composite list now averages about 3 billion each day.) When it became apparent that the strain was not about to let up, the stock exchange closed every Wednesday, giving bookkeepers time to update accounts before the next day's activity brought a new pile of paperwork. In January 1969, the exchange resumed five-day trading, but closed at 1:30 p.m. instead of 3 p.m. to accommodate the crush of paper. Control Data Corporation, it turns out, was uniquely prepared to benefit from this crisis. Its medium-range computers had appropriate software to handle the problem, and the company received a sudden influx of orders from Wall Street firms forced to abandon their old paper-based ways.

In 1968, the Minnesota Securities Department received 893 first-time applications for securities registration, obviously a heavy workload. The commissioner determined that none of the state's four examiners were being precise enough in their examinations and suspended them for a week without pay. His action nearly shut down the beleaguered department. For the year, an estimated one in twenty applications was approved.

Perhaps the lesson to be drawn from this feverish era is that it's helpful to encourage businesses financing whenever possible and to expect an eager speculative reaction, knowing that it's not easy to differentiate between the successes and the failures ahead of time. Maybe the menial worker on the steps of City Hall eventually lost all his money—there was no follow-on report—but he had a rush of success, and he was helping to support a market that brought a lot of worthwhile local companies into prominence.

By 1970, stock market volume had returned to normal and trading was once again open a full five days a week. Control Data computers had become an indispensible part of the new standard of operation.

8

Success and Hard Times

"Paper machines," Norris called them; that is, computers that were not yet fully developed and therefore not deliverable. He was convinced that his principal competitor, IBM, had announced the availability of such machines in 1964, well before they were ready to be manufactured and shipped, in order to forestall customer decisions to buy Control Data supercomputers. Because of IBM's size and influence in the industry, purchasing agents had to consider its offerings when planning a procurement, and then make a competitive evaluation. The doubt and confusion created by the IBM announcements helped drop Control Data into a loss position in 1966. Norris was convinced the IBM tactics were illegal.

Intent on making his case, Norris set about gathering evidence. He assigned Jim Miles, one of his senior marketing executives, to manage the effort together with legal counsel from the Oppenheimer law firm in St. Paul. They gathered IBM public announcements, general press articles, reports from company sales representatives, and interviews with select customers; to prove "monopoly" power, they measured the size of the total market. The reports attributed "70 to 80 percent of the total computer systems installed in the United States to IBM, with the balance held by other competitors, not one of whom accounted for more than eight percent." The accumulated evidence showed that IBM had offered machines at unprofitable prices, and to compete, Control Data had to agree to terms it couldn't afford, which resulted in losses. The tactics were aimed specifically at the Control Data 6600 system, the undisputed leader in the industry. It was, Control Data believed, an abuse of monopoly power. Various summary reports were compiled by the company and passed along to the U.S. Department of Justice, Antitrust Division, making the case that IBM was deliberately attempting to monopolize the industry. By supplying evidence, Control Data was inviting the government to enforce the Sherman Antitrust Act and bring action against IBM.

Over the next two years, Control Data filed additional reports and the Justice Department began to gather information. Congressman Emanuel Celler of the House Antitrust Subcommittee conducted hearings exploring monopoly conditions in the computer industry, and all manufacturers, including IBM and Control Data, were required to supply data. But the various investigations resulted in no action.

Its initial objective unmet, Control Data was then faced with a decision. If it brought charges on its own, the company was required to do so within four years of the infractions, and the date was rapidly approaching. Further, the company was about to announce the completion of Seymour Cray's latest supercomputer, the 7600, and it didn't want a repeat of the tactics IBM had used to undermine sales of the 6600. But to proceed alone could result in a "long fruitless, expensive and potentially disastrous course," warned one of the Control Data attorneys. It could become "a legal battle of huge proportions in which [Control Data] would be out-manned, out-resourced, discovered-to-death, and ultimately overwhelmed by a vastly superior evidence-gathering machine." Surely the decision was not an easy one to make. The company's directors questioned the wisdom of the action and asked for additional opinions from other national law firms with known antitrust experience. Finally, facing an approaching deadline and the possibility of abandoning years of work gathering evidence, the company decided to proceed on its own. On December 11, 1968, Elmer Trousdale and his partner, John Robertson, both Oppenheimer attorneys, walked over to the federal district court and filed an antitrust complaint against International Business Machines Corporation on behalf of their client, Control Data Corporation. With that, the company became David taking a slingshot to Goliath. It stood alone in challenging the world's largest and most successful computer manufacturer.

Having filed suit, Control Data was now in a position to pursue discovery of IBM's own records for evidence supporting its case, and IBM was in a position to do the same of Control Data. Despite the "surprise" element of the case, within ten days IBM had "served a massive set of interrogatories."

Three weeks after the Control Data filing, a company called Data Processing Financial and General filed suit against IBM, charging monopolization of computer leasing. And two weeks after that, on the last day of the Johnson administration, January 17, 1969, the Justice Department filed suit charging IBM with monopolization of the computer industry. Richard Nixon from the opposition Republican Party had defeated Democrat Hubert Humphrey and was about to take office. Whether that was a significant consideration is not publicly known, but the suit was filed by the outgoing administration on the last possible day. Control Data now had confidence that others, too, felt that the evidence existed to convict IBM. The company then served IBM with its own set of "massive interrogatories."

Other parties filed suit against IBM in the spring of 1969, and still another filed in the following year, but in the opinion of Control Data's lawyers these suits hindered the company's case because IBM used them to promote delay and then filed a motion to consolidate all cases. A judicial panel agreed to the consolidation, and IBM was able to argue that any document or interrogatory answers it provided must come from a coordinated request of all parties. This, of course, added time and expense to the plaintiffs' efforts and made coordinating difficult, as some of the parties were concerned about computer leasing, some about software, and some about hardware market dominance. Control Data counsel also felt sure the cases would be transferred to New York, adding further expense and complication to their work. However, in July 1969, the cases were assigned to U.S. District Judge Philip Neville in St. Paul.

When the discovery process got underway, IBM resisted producing various documents by claiming "privilege" and lack of "good cause," particularly the material showing its competitive market position, and also resisted supplying the material given earlier to Representative Celler. IBM was required to deliver such documents, however, and they showed that IBM held a dominant market position, confirming the earlier submissions by Control Data. These documents helped establish the position of "market power."

The discovery process also produced other documents, helping to support Control Data's case. One referred to IBM's "price control and its supporting practices." From President Tom Watson's files came a copy of the Control Data complaint alleging IBM's competitive moves to thwart competition, and his handwritten comment: "Probably not true, except for the model 91," which was the high-end model directly competitive with the Control Data 6600. There was also an internal document saying the company "announced some of the machines 24 months early, and the total line an average of 12 months early." From executive T. V. Learson: "We put a delivery date on something yet to be invented." Other internal memos showed projected losses on the high-end systems, followed by price reductions that would increase the losses. The information supported Control Data's basic argument that IBM was attempting to monopolize the market.

Another document involved IBM's effort to settle the case with the Justice Department, including the terms, indicating that IBM would consider various alternatives, possibly less damaging than those proposed by the Justice Department. Meanwhile, Control Data itself was producing "80,000,000 documents" for IBM. (That's 80 million, the number taken from a summary of the case written by attorney Elmer Trousdale. It could be a mistake in zeros, but in any event it was a large number.)

In 1970, IBM reached early settlements with various contesting parties, and by the end of that year the only remaining cases were Control Data, Greyhound Computer (a leasing company), and Telex, the former hearing aid company that had moved from Minneapolis to Tulsa and was now producing computer peripheral equipment. Late in the year, IBM's chief counsel, Tom Barr, invited Control Data lawyers to evening dinner in White Plains, New York, intending, as he said, to "smooth things over." After a pre-dinner round of martinis, he raised what sounded like a settlement overture: "What does Control Data want in this case?" Nothing came of the discussion, but the Control Data lawyers did note the implication of the question.

By late January 1971, Control Data had completed its review of IBM documents. It included a telling array of admissions: that IBM had announced computer models with the knowledge that they would lose money; that IBM had announced models prematurely to thwart competition; that IBM had effective control over industry pricing; and that the company had pursued a strategy to increase market share.

But the defendant had made its own discoveries and counterclaimed that Control Data had coerced customers to buy equipment; along with Commercial Credit had restrained trade through reciprocal business arrangements; and had restrained trade by way of its numerous other acquisitions. Months were spent skirmishing over these points. It was expensive and stressful. Eventually, Judge Neville issued a ruling denying many of the IBM counterclaims. The Control Data lawyers felt a sense of relief and gained renewed confidence. But one month later, on July 7, 1971, the company received a shock nobody imagined would happen. John Robertson, one of the Oppenheimer attorneys handling the case for Control Data, suffered a massive heart attack and died. He had provided direction and planning for the case, and was the principal legal representative in all substantive discussions with IBM. Control Data had elected him to its board of directors the previous November. He was just forty-eight years old. It was a stunning development, but because he was one of a team, the lawyers regrouped and were able to proceed with the case.

In August 1971, IBM asked for court approval to take a census of 2,700 companies to properly define the market. It was to their advantage to broaden the definition as much as possible, and they intended to include leasing companies and companies engaged in the resale market. Of course, the effort would also further delay the case. Judge Neville approved the request. Control Data in the meantime had refined its numbers to indicate that in the early 1960s, the relevant period, IBM had about 75 percent market share, while Control Data had 4.5 percent. Considering only machines sold to the U.S. Government, IBM had about 50 percent and Control Data 9percent of the market. In either case, according to the Control Data numbers, IBM's dominant market position was clear.

In November 1971, Control Data filed a pretrial memorandum stating "factual and legal contentions" about which IBM was supposed to respond. Control Data lawyers thought the process would "narrow the issues in dispute and hopefully shorten the trial." It did not happen. Instead, IBM provided evasive answers or gave meaningless responses using the code "DKI," meaning denying knowledge and information of the point involved. Sometimes their answers simply contradicted one of their own documents. Judge Neville ordered them to be more responsive. In late 1971, depositions also started. Throughout the next year, "about 60 depositions of IBM executives were taken." Each involved thorough preparation by the Oppenheimer attorneys, followed by a flight to IBM offices in New York for an indefinite stay. The lawyers later commented that the depositions were "simply remarkable," saying that the memories of IBM executives went strangely blank and they were unwilling to acknowledge even simple facts. The Control Data lawyers came to believe that the witnesses had been "carefully coached." One witness gave his age and then withdrew his answer saying he didn't know his age, that he must have been told his age and had no particular knowledge of it. Another witness testified that he didn't remember what he meant in a memo he had written or why he wrote it. Another was asked if he had written a memo that indicated that he was the author, and he said "no," meaning his secretary had written (typed) it. The questioners began to worry that a witness might deny his own signature on a piece of paper because the paper was a copy, rather than an original, and felt the need to contort their questions meticulously to prevent such creative evasions. The head of IBM's Systems Development Division claimed to have never heard of the Model 90 series, even though his division had supposedly developed it and he had signed many documents mentioning it. In the end, the Control Data lawyers concluded that "as a general matter, the depositions were a farce." Attorney Elmer Trousdale said that he deposed "about fifteen IBM officials," including top officers, and concluded: "Not a single useful admission was obtained from any of them, despite the damaging accumulation of IBM's documents arrayed against them." He later described the effort as "tiring, frustrating, and seemingly unending work: first, to face mountains of mostly useless documents; and then, the know-nothing, pettifogging, IBM witnesses," and he was entering his third year of trips to Armonk, north of New York City. He added an afterthought: "We should have suspected that IBM had no intention of ever trying Control Data's case."

Although Control Data and IBM were thrusting and parrying in Judge Neville's chambers, serious discussion about settlement had begun in May 1971. After the death of John Robertson in July, attorney Dick Lareau of the Oppenheimer firm took the lead for Control Data.

Early in the negotiations, Control Data proposed divestiture of various IBM divisions. IBM advised that it was not prepared to discuss that subject with Control Data. Without abandoning divestiture, Control Data also proposed that several IBM practices be prohibited by injunction. IBM responded that it would agree to properly phrased restrictions on its practices as part of resolving both the Control Data and the Justice Department cases. By December 1971, the following terms evolved.

1. IBM would pay Control Data its legal fees on condition Control Data would give IBM and no one else all pretrial work product.

2. Ownership in IBM's Service Bureau Corporation would be turned over to Control Data for payment at book value in cash or stock. IBM would guarantee not to compete for a limited period, and IBM would agree to purchase a negotiated level of services over a five year period.

3. IBM would purchase from Control Data about $4.5 million worth of disc packs.

4. Control Data would provide to IBM various development services involving peripheral products.

Almost a year went by as the various points were discussed, and then on December 8, 1972, Norris and IBM executive T. V. Learson had a clandestine meeting in Omaha, Nebraska. Norris asked Learson to value the proposed IBM settlement package. The Service Bureau Corporation earned $5 million, Learson said (he probably meant before taxes and other charges), and applying a multiple of 15 times earnings created a value of $75 million. Together with the other considerations, he placed a value on the total package at $150 million. Norris disputed the valuation and rejected the offer. They agreed to meet again.

Learson then offered $26 million in benefits for Service Bureau employees and a paid-up patent license for five years. On December 21, the two met again and resolved various issues. The next day, Learson advised Norris by letter that his latest offer was final. Norris apparently decided that he had pressed the issue long enough, and the lawyers began to work on many underlying details of the proposed offer. On January 11, 1973, a settlement was announced. If Control Data would destroy its data base of pretrial work product, not the evidence itself, but the method of storage and retrieval, IBM would hand over checks in the amount of $66 million plus all of the common stock of the Service Bureau Corporation.

The next day, Control Data was able to certify that the data base was dismantled. IBM delivered the checks, and Judge Neville signed a one-line order of dismissal.

A summation of the financial terms of the settlement could be stated as follows:

Value of Service Bureau Corporation	$75 million
Less purchase price	- 16
Net value of Service Bureau Corportation	59
IBM purchase of services	25
IBM payment of fringe benefits	+26
IBM payment of legal fees	+ 15
Subtotal	66
Other IBM purchases over 5 years	30
TOTAL	$155 million

However, several newspaper articles valued the Service Bureau Corporation at $50 million to $180 million, so there was subjectivity to the valuation. Of course there were also the intangible benefits of the settlement. These would include the tacit agreement that IBM had indeed engaged in behavior as charged and was eager to settle even to the point of paying lawyer fees; the prestige of winning against an immense foe and the respect this would bring in the marketplace; and, not insignificantly, the restraint now placed on IBM's future competitive behavior. The case was finally finished after four years of legal wrangling.

Whether Norris was wise in accepting IBM's settlement terms rather than going to trial cannot be answered with certainty, but subsequent events suggest he made the right choice.

Judge Neville had issued an order sending the Greyhound case to Arizona for trial. Greyhound tried the case before a jury in Phoenix, and the court gave a directed verdict in IBM's favor. This appeared to be an IBM victory, but it was a temporary one, as the decision was later reversed on appeal.

The Telex case was moved to Tulsa, Oklahoma, and in the spring of 1973 the judge issued a decision ordering IBM to pay $353 million to Telex, minus $22 million in damages Telex was ordered to pay for stealing IBM's trade secrets. Telex appeared to have won big, but that decision was reversed on appeal, except for the $22 million Telex was ordered to pay. Telex, then faced with bankruptcy, eventually reached a negotiated settlement with its adversary.

The zig-zag direction of these decisions gave no indication of Control Data's chances, although the attorneys handling the case on Control Data's behalf were confident they had compelling evidence in the supercomputer end of the market, and they had the reinforcing support of the Justice Department. But Justice contributed very little. Their attorneys attended many of the IBM depositions mostly as listeners, and they relied primarily on Control Data for the production of documents. They did, however, show an interest in providing injunctive relief against IBM practices and, after conferring with members of the industry, considered splitting IBM into three pieces. Control Data attorneys, fearing that three new entities, each still much larger than any other rival, would not be a satisfactory result, instead urged divestiture into "many, many pieces or none at all." Ongoing discussions produced a variety of divestiture possibilities. Nothing came of them. Nine years later the Justice Department abandoned its case, declaring it "without merit." In summation, a Justice spokesman noted that even if there was evidence of predatory pricing, which seemed likely, it did not lead to monopoly power since Control Data went on to significant commercial success, selling approximately ninety five of its 6600 supercomputer systems.

In hindsight, Control Data was indeed fortunate in bringing its case alone, for it gained a settlement that might not have been otherwise possible. The Service Bureau Company, a business with about forty offices throughout the country, selling computer *time* to many customers unwilling to make the investment in their own machines, had became a substantial business. It had a reliable stream of revenue that was independent of the many extraneous factors affecting the production and sale of computer hardware, and it offered the opportunity of selling computers to its customers once they reached a point where it was economically feasible. The business became a fourth leg to Control Data's three other principal activities: the production and sale of computers, the production and sale of peripheral hardware, and the financial services of Commercial Credit. From all appearances, Control Data now had a strong foundation, benefiting from multiple sources of revenue, and was poised for balanced success.

One of the subsidiary issues in the antitrust case was the industry method of price "bundling" as practiced by IBM and therefore others out of competitive necessity. A customer was charged a single price for hardware, software, programming support services, training services, and maintenance service. The separate activities were not identified as individual components in the overall price. Such an arrangement made it very difficult for a competitive supplier to charge for separate software or maintenance when it was seemingly "free" from the supplier selling a "total system." In effect, the practice gave IBM total control over its customers, and therefore control over approximately three-fourths of the industry. Only in June 1969, after the antitrust lawsuit began, did IBM announce separate charges for application software, but the remainder of its products and services were still bundled within the single offering price.

Norris, of course, had been building a diversified company both through acquisitions and internal development. His peripheral products division was selling printers, tape units, disc drives, and display terminals to other computer manufacturers who would then resell the products under their own names. Such products could also be sold by Control Data directly to customers of IBM if they had a basis of comparison. The same could be said for maintenance services. He had also started a training division willing to serve anyone in the industry. All of these offerings had a difficult time gaining entry to IBM accounts, regardless of their competitive advantages.

To counter this, Norris devised a strategy to open up the market for these products and services—particularly to the customer base of IBM—by separately identifying and pricing all hardware products, software products, and services delivered with Control Data computer systems. Such itemization would give customers a greater range of choices and more control over their buying decisions. As software and services assumed a larger portion of total system sales—which Norris foresaw happening—the customer would be able to allocate resources to those products providing greater benefit. Further, Norris maintained that the competition brought about by unbundled services would result in a higher quality of offerings for customers. From Control Data's business standpoint, unbundling would provide new market opportunities, especially within the IBM base of business. So confident was Norris of this approach that the cover of the annual report for the year ending December 31, 1969, featured a bold pronouncement.

> *There have been many important events as the computer industry has progressed through its 20 years of history, but there are only two which are truly milestones: the change from vacuum tubes to the transistor, and now separate pricing—unbundling.*
>
> – *William C. Norris*

His leadership in this was significant, and the history that followed proved him right, as whole industry groups formed around plug-compatible peripherals, third-party maintenance, programming services, and software packages. The customers benefited and so did the many new companies that sprang up to provide these products and services. In many ways, unbundling opened up the industry. But it also brought price competition. The IBM "umbrella" of prices over the industry soon lay in shreds, affording little in the way of protection for those competing beneath. Many of the new smaller companies selling hardware or maintenance services, having to compete in price against one another, eventually either consolidated or faded away. Japanese companies entered the disk drive market and created intense price competition. Control Data suffered along with other companies from the dog-fight conditions that developed in the various hardware offerings. But, in software, the change created an opportunity for companies like Microsoft and Oracle to flourish. The strategy of "unbundling" changed the industry in many ways, and Norris's bold leadership helped bring it about.

While companies were going about their business and stocks reached new highs in 1968, the United States as a country was undergoing profound social upheaval. Many youth considered the war in Viet Nam an unwarranted American intervention, burned their government draft cards, and marched in the streets in protest. Some fled to Canada, refusing to serve in the military. A few entered government offices and poured blood over draft files. At the same time, America's black citizens demanded an end to segregation and discrimination in both the southern and northern parts of the country. They wanted the same fair treatment as other citizens, the right to use all public accommodations, the ability to move into the neighborhood of their choice, the opportunity for a decent job. Black anger and frustration grew so intense during what became known as "the long hot summer" of 1967 that many inner-city neighborhoods were looted and set on fire. The following spring, black civil rights leader Martin Luther King was assassinated. Presidential hopeful Robert Kennedy was also assassinated. Cesar Chavez was making demands for migrant workers in California, American Indians found a new voice in an organization called AIM, and feminists were calling for true equality for women.

But beneath all the turmoil, much of American life went on as always. Businesses for the most part ignored the social discontent. What did civil rights and political arguments over a war and poverty have to do with the country's commerce? Companies continued to focus on their work.

Norris, on the other hand, was impelled to act. His business was successful and growing, and he believed it could be a larger force for good. Many of the orders for his company's products came from agencies of the government. If he addressed social problems that were vexing the government, maybe there would be reciprocal benefit. He never described his motivation that way, but in the annual report of 1968, while summarizing the year's activities, he included a paragraph titled "Participating in solving poverty problems," and explained that his company had opened a plant on the north side of Minneapolis—a largely black, low-income area—and was planning another facility in a low-income area of Washington D.C. These plants were efforts by the company to locate and train potentially skilled employees, he said. It was the beginning of a significant new emphasis by Control Data to address society's problems while continuing to meet the needs of its customers, employees, and shareholders. Like so many of his other initiatives, it was an experiment. His detractors doubted its effectiveness, but he proceeded anyway. He saw it as a business opportunity and felt he had the resources to capitalize on it.

One former high-level executive who preferred to remain anonymous disagreed with Norris's approach but was willing to be quoted on the man's personal strengths. "We're talking about a very complex character. He is a highly intelligent, hardworking guy. He doesn't give a damn about money. He wears old suits. He drives old cars. I used to think it was an affectation, but it is genuine. He looks like Carl Sandburg; he acts like Carl Sandburg. He's a farm boy. He doesn't want anything to do with people from New York. He has a dislike for securities analysts and MBAs." Further, the executive might have added, there was little in Norris's rural background to suggest to outsiders an understanding of inner-city issues or an interest or willingness to address them. It was a surprising impulse. Some have suggested it came from seeing his family and neighbors suffer hard times during the Depression and the assistance they received from Franklin Roosevelt's reforms. Perhaps he saw himself as the sheriff of a frontier town, but in a different setting, standing for what is right. He seemed to take pride in the singular role. Earlier in his career he once said, "When I see everybody going south, I have a great compulsion to go north." He sensed, or knew, that originality begets opportunity.

In September 1969, the company opened a plant in the impoverished Appalachian area of Eastern Kentucky, and later that year opened an electronics sub-assembly plant in a disadvantaged area of inner-city St. Paul, Minnesota. The purpose of these plants, Norris said in his year-end report, was to bring together Control Data's need for people and those communities' need for jobs. The Kentucky plant was in a county with the second-lowest per capita income in the country. His initiatives were an attempt to blend high-minded social values into the machinery of business.

But while Norris was launching these new initiatives, the prosperous core of his business again dropped into recession. There were slippages in order closings in both computer hardware and peripheral products, and in some cases, because of the economy, there were outright order cancellations. Some equipment on lease was cancelled and returned. The lawsuit against IBM was still underway, and Norris referred to it in saying there was "intensified marketing pressure from our major competitor." Internationally, however, things were better. Volkswagen had placed an order for multiple computers valued at $14 million, and CERN, the European organization for nuclear research, had placed an order for the company's new 7600 supercomputer. Nevertheless, when the year came to a close on December 31, 1970, the company reported total computer revenue, including peripherals and services, of just $550 million, down from $578 in the preceding year, the first such decline in the company's history. Substantial inventory write-offs, plus provisions for actual and anticipated excess costs on a number of government contracts, helped cause a loss from computer operations of $41 million. The Commercial Credit subsidiary was able to increase earnings approximately 10 percent from a year earlier, achieving a profit of $38 million, but not enough to offset the computer-related losses. The overall corporate result was a loss of 34 cents per share, down from earnings of $3.62 a share in the preceding year. It was the second time in five years that the company had fallen into a loss position. All computer business employees were given a 10 percent reduction in work time, and therefore reduced pay, during the last five months of the year. Norris ended his remarks in the annual report by saying, "Our computer business is budgeting for growth and a small profit in 1971," and the "financial services business [Commercial Credit] is looking forward to another good year." The stock market made its own evaluation, however. The image of a reliable growth company had been shattered, and the stock traded as low as $29 that year.

Contrary to the budgetary expectation, the computer operations lost money again in 1971 and once again in 1974. In 1975, "salaries for top corporate management and many higher level employees [were] set at reduced levels, [and] salary reviews [were] delayed for most computer business managerial and professional employees." In spite of its success in settling the IBM lawsuit in early 1973 and the additional revenue and earnings brought in by the newly acquired Service Bureau Corporation, it wouldn't be until 1979 that earnings from computer operations would exceed the earnings of financial services. Thus, for nearly a decade the bear was carrying the whippet, and all the earlier assumptions about Control Data's high-growth operations easily outpacing the stolid activity of Commercial Credit proved inaccurate. In 1974, the stock hit $9½. Overall corporate earnings in that year were 13 cents a share, achievable only because Commercial Credit was able to contribute enough to offset the loss from computer operations.

Nevertheless, the strategy of addressing social concerns as a business opportunity continued. In 1975, Commercial Credit bought and restored ten row-houses scheduled for demolition and offered them for sale to low- and moderate-income families. Control Data assisted in training twenty health care representatives from the Rosebud Sioux tribe in South Dakota and placed a mobile clinic with Control Data terminals in operation on the reservation. Overall, the company made a concerted effort at affirmative action "to increase the quantity and quality of job opportunities for minorities and women." The new emphasis on a broader social outlook was suggested in a statement on the cover of the 1977 annual report. "Control Data is more than a computer company. Over the years we've developed a unique business strategy—combining computing technology, financial services and industry expertise...[thereby] helping to improve productivity and the quality of life throughout the world." It was visionary; it was difficult; and it was almost a complete change in corporate personality from the earlier supercomputer hardware company that left customers with the task of programming their own computers.

By 1972, Seymour Cray had resigned from the company in all capacities, although the press release said he was "phasing out." For those who understood his technical importance to the company, it was appalling news. Trading in Control Data stock was briefly suspended. Seymour explained later that he had been given all the support he needed from Bill Norris, and recalled somewhat nostalgically that when he originally set up in Chippewa Falls, Norris told him, "Just do whatever you want to do, and we'll take care of it." But then the company got so big, he said, and "there were all these middle management types strategizing and making market plans." Further, there was a competing development effort within the company for a large computer, called the STAR-100. "There really had to be a separation," he said. "A personal relationship [with Bill Norris] wouldn't quite do it."

Seymour would remain in Chippewa Falls, Wisconsin, form his own company, and design a new line of supercomputers. Frank Mullaney, his earlier boss at Control Data, would join him as an investor and become chairman of the new company. Earlier in his career, Seymour had said that he liked to begin each project with a clean sheet of paper. Now he was able to do that, just as he'd done fifteen years earlier when Control Data began. He would not worry about compatibility with earlier product lines, or expansive software applications, or the requirements of existing peripheral products, or bothersome requests from corporate staff. He would design the best, fastest computer he knew how, and he was confident there would be customers for his new product. His company would be called Cray Research, a title simple enough to convey its purpose, for Seymour's reputation was then widely understood and appreciated. But once he left to form his own company, Control Data lost forever its reputation as the world's leading designer of supercomputers.

Though its impact was profound, Seymour Cray's departure wasn't the only factor affecting the deterioration of Control Data stock during that difficult time. There were larger national influences that had an effect on virtually all stocks across America. On March 17, 1973, the Watergate committee of the U.S. Congress began hearings that implicated President Nixon and others in his administration in a politically motivated burglary. The investigation soon uncovered a whole range of abuses concerning the misuse of agencies of the federal government, and the news media across the country reported them in an unfolding drama. On April 30, two top White House officials, H. R. Haldeman and John Erlichman, resigned, as did Attorney General Richard Kleindienst. The same day, White House counsel to the president John Dean was fired. On October 10, Vice President Spiro Agnew resigned on money corruption charges. Eleven days later, on a Saturday, President Nixon arranged the firing of special prosecutor Archibald Cox and accepted the resignation of Attorney General Elliot Richardson and Deputy Attorney general William Ruckelshaus, who refused to carry out his orders. These actions drew sensational headlines.

Compounding the turmoil, in October 1973 the Arab oil-producing nations announced they would cut back on oil exports to Western nations and Japan. Among other things, they were upset by the outbreak of an Arab-Israeli war on October 6, the fourth such conflict in twenty-five years. Although the war lasted just three weeks, the embargo lasted six months and caused world oil prices to quadruple. The United States experienced gas shortages. In an effort to reduce consumption, Congress passed, and President Nixon signed, a bill imposing a 55-mile-per-hour speed limit for those states accepting federal highway funds (1974 Emergency Highway Energy Conservation Act). More importantly, the dramatic increase in the cost of a commodity so fundamental to the economies of the developed world caused wrenching repercussions in virtually every business. Prices were raised on many items in order to cover increased costs, while at the same time companies adjusted their activities wherever possible to reduce their dependence on petroleum.

The Viet Nam war was still underway, and its financial cost contributed to the accelerating inflation. With inflation comes higher interest rates, and the prime rate rose from 4.5 percent in February 1972 to 12percent in July 1974. Borrowers were forced to adjust quickly to this near tripling of costs. The change, of course, had a negative effect on the earnings of Commercial Credit, a regular borrower in the capital markets, and earnings dropped in half over the three-year period from 1972 to 1975. There was a similar effect on all borrowers.

After lengthy and at times dramatic testimony, the House Judiciary Committee passed articles of impeachment on the president of the United States. On July 27, 1974, acknowledging the momentum against him, President Nixon resigned. American business, while not directly affected, was working under a Washington political quagmire. Though it had traded over 1000 in 1972 and 1973, the Dow Jones Industrial Average hit a low of 570 on December 8, 1974.

Control Data stock took a swan dive during this period along with many other firms. Below is a listing of various companies in the Twin Cities and their respective declines in price. It was not a good time for investors anywhere, except for a savings account earning a high rate of interest.

	High (Year)	Low (Year)
Applebaum's Food Markets	31.5 (1971)	1.0 (1975)
Arctic Ent.	45.4 (1971)	1.2 (1975)
Cardiac Pacemakers	26.0 (1974)	9.0 (1974)
Cherne Indusrial	20.2 (1972)	1.0 (1974)
Coca-Cola Bottling	41.2 (1972)	4.8 (1974)
Comten, Inc.	50.0 (1970)	1.4 (1974)
K-tel Int.	28.0 (1972)	1.1 (1974)
LaMaur Inc.	40.0 (1970)	2.2 (1974)
Leisure Dynamics	20.0 (1970)	.3 (1974)
McQuay Perfex	40.6 (1969)	5.6 (1974)
Medical Investment	20.0 (1969)	.4 (1974)
Medtronic	66.0 (1973)	19.4 (1974/5)
Minnesota Fabrics	42.2 (1972)	1.7 (1974)
Minnetonka Labs.	42.2 (1972)	.4 (1974)
Modern Merchandising	37.2 (1972)	1.3 (1974)
Munsingwear	38.6 (1972)	9.6 (1974)
Pacific Gamble Robinson	47.6 (1972)	16.0 (1974)
Possis Corp.	124.0 (1966)	1.6 (1974/5)
Sci-Med Life Syst.	19.0 (1974)	3.3 (1975)
Shelter Corp.	14.5 (1972)	.1 (1974)
Standard Fabrics	16.0 (1970)	.2 (1974)
Sterner Lighting	29.2 (1971)	1.0 (1974)
Tonka Corp.	29.0 (1970)	5.2 (1974)

Toro Co.	51.0 (1969)	8.0 (1975)
Turbodyne Co.	25.7 (1971)	2.1 (1974)

Someone owning all of these stocks in a balanced portfolio (like an index fund today) probably thought he had sufficient diversification to withstand serious market disruption. It was a reasonable thought, but wrong. Overall, the portfolio would have dropped a drastic 90 percent in value, although in truth it would have been unusual to have bought every stock at the absolute top. Nevertheless, the comparison illustrates the trap-door collapse of stock prices at the time.

Even the big quality companies were affected. Dayton Hudson (Target) fell from 39.6 in 1971 to 6.4 in both 1974 and 1975, a drop of 84 percent, even though the company earned $1.70 in 1974 and $1.57 in 1975 and paid a dividend of $.54 a share. At its low, the stock was selling at 4 times earnings and provided a dividend yield of 8.4 percent.

Minneapolis Honeywell stock fell almost 90 percent from 170.8 in 1972 to 17.5 in 1974. In comparison, the highly respected growth company across the river, Minnesota Mining and Manufacturing, benefited from its reputation of financial consistency and dropped only 53 percent, from 91.6 in 1973 to 43.0 in 1975.

Northwest Airlines fell from 55 in 1972 to 10.5 two years later. Since it earned $3.00 a share in 1974, it was selling at approximately 3 times earnings.

Northwestern National Life Insurance fell from 27.3 in 1972 to 7.6 in 1974 and at its low was also selling at 3 times earnings. One would think an insurance company would be more stable than an airline and worth a higher multiple. Such thinking would be justifiable, but also wrong. This was no ordinary time.

St. Paul (insurance) Company dropped from 55 to 15 in two years. Probably because its financial history going back to

1853 inspired trust, the stock never got below a comparatively expensive 8 times earnings.

Because interest rates had risen so much during the period, utility stocks were also affected. In a publicly traded market, the dividend yields must compete against other relatively safe returns, such as a savings account. If the dividend stays the same, the price will drop until the yield compares satisfactorily. Northern States Power stock fell in half from 1973 to 1974 and at its low provided a yield of 12 percent. Ottertail Power fell only 39 percent and offered a yield of 11 percent. For the same reason, a similar adjustment happened to bonds, which, of course, are considered safer than stocks.

It was a bear market, that's for sure. From an inflated bull market based on ever-expansive earnings expectations, a rash of new issues, and continuing buoyant prices for more seasoned stocks, the mood darkened to a complete lack of confidence in any stock, almost regardless of quality; and there was a repricing of even the safest stocks in order to provide a higher dividend yield. Yes sir, it was a bear market. Thank you, President Nixon. Of course, it wasn't entirely his fault.

A report came out of the Minneapolis investment community citing the carnage in local prices and offered a telling suggestion. When your parents talk about the Great Depression, it said, and tell you about the money lost, tell them you know pretty well how they felt. From 1929 until 1932, when President Roosevelt was elected, the Dow Jones Industrial average dropped 89 percent. Radio Corporation of America (RCA) was probably the Control Data of its day, representing not computers but the new medium of radios. It had started in business in 1919. Two years later, the stock was quoted at 1.4. By 1928, it was a sensation, blazing in price from 85 to 420 during the first eleven months of the year, and then, suffering a fear of heights, dropped 72 points in one day during December when there was a sudden break in confidence. That jarring adjustment was, perhaps, a sneak look at what might eventually follow. But in February of 1929 the stock split 5 for 1, and on September 3 of that year, "Radio Corporation of America, adjusted for earlier split-ups and still not having paid a dividend was 505," according to *The Great Crash, 1929*, a classic history of that time. Other sources tell us the stock was selling at 72 times earnings at its peak and that it lost 97 percent of its value by 1932.

Learning nothing from the investment manias of earlier times, each generation seems to take its turn. Author John Kenneth Galbraith called it "a basic recurrent process," and in explaining the crash of 1929 said that "in the U.S. in the nineteenth century there was a speculative splurge every twenty or thirty years."

In spite of the lingering headache that follows an investment binge, the spirit of innovation remained, and even though most stocks were caught in a sinkhole in the local market of the mid-1970s, a new medical industry was quietly taking shape.

9

The Entrepreneur

If William Norris helped create the computer industry and was smart enough to nurture the genius of Seymour Cray, and if Earl Bakken created a medical device industry that no one else could foresee while soldering parts at his workbench, then Manuel Villafana was simply a man looking for opportunities he could bring to fruition in the financial afterglow of those companies. He was a natural *entrepreneur*. Capitalizing on the public's willingness to finance unproven ideas, he started an astonishing seven medical companies. Now in his seventies, Villafana is working to nurture his current enterprise, Kips Bay Medical, to full health. Like Seymour Cray, he came to be known by his first name only, Manny—a name unique in local investment circles and seemingly recognized by all. His business aptitude, although more general in nature, was just as brilliant as the others'. Not all of his companies scored big, but some certainly did. Very big.

Manny's parents, both Puerto Rican, moved up to New York City shortly before the stock market crash of 1929. The family struggled more than most during the depression years that followed, as employment was scarce for a man with no education and limited English. There were times when there was no food on the table, an older son remembered. Manny was born on August 30, 1940, the last of four boys. His mother was forty-eight years old then and his father fifty-nine, which made them, by age, almost his grandparents.

When World War II began, members of the family found employment in the defense industry in nearby Connecticut. But after the war, Manny's mother moved with the two youngest sons back to New York City into the Spanish Harlem neighborhood of the South Bronx. Two years later, her husband died, a week before planning to rejoin his family, worn down by the difficulties of finding his place in an unfamiliar land and the long-term effects of heavy drinking. Manny was nine years old at the time, almost an only child, as his older brothers had by then gone out on their own. His mother worked as a seamstress. He was left to find his way during the daytime in a neighborhood suffering from poverty and crime.

One day he asked a friend where he'd been and learned of the Kips Bay Boys Club, a subway ride away on the east side of Manhatten. It had been set up in 1915 by businessmen for the purpose of keeping young boys off the streets. There were games, athletic events, swimming, woodworking, and other activities. Manny joined the science club and found that he enjoyed taking things apart to see how they worked. He also got a job taking care of the gym equipment and earned 40 cents an hour, money that helped at home. Some of the men working at the club took an interest in him and became, by his own description, surrogate fathers. They kept him busy and away from the temptations of the city that might lead to trouble. He became captain of a ping-pong team that won the city championship and went on to compete against international competition in the U.S. Open. He was also captain of the swimming team. Kips Bay, he said, saved his youth.

By high-school age, Manny had saved enough money to pay his own way to Cardinal Hayes Catholic School in the South Bronx. There were four thousand students, all boys, but it was a quality school with rigorous discipline and expectations of superior achievement. Manny received good grades in math and science and perfected his English, as Spanish had been his first language. He grew into a tall, fair-skinned man with a ready smile.

Manny Villafana, seventeen, with his mother, Elisa, in their apartment in South Bronx

After graduating from high school, he enrolled at Manhattan College, but ran out of money after one year and dropped out to take his first full time job proofreading for Radio Engineering Labs. He also went to night school to learn electronics and soon became a technician, something that came to him quite naturally. At age twenty-three, he went to work for Ethyl Corp., where he learned general management skills as an assistant business manager running a small gasoline testing lab.

Manny's next position was at Picker International, a unit of Picker X-ray, working as a customer service representative handling medical products. Again expanding his repertoire of knowledge and skills, he coordinated suppliers with customers and when there were problems spoke from his office in New York to doctors all over the world. Medtronic became one of his accounts, and pacemakers one of the products he had to explain. At the time, pacemakers often failed in twelve to twenty-four months, generally because of depleted batteries; the patients didn't die, but returned to their pre-implant lethargy. Trying to learn all he could about the Medtronic product and at the same time build rapport with members of the company, Manny traveled to the Twin Cities a few times. On one occasion he spoke with Bill Greatbatch, the inventor of the Medtronic implantable, then giving a seminar in Minneapolis, and made an acquaintance that would benefit him greatly as his career unfolded.

Two years later, Medtronic decided not to renew its agreement with Picker, the intermediary to its international accounts. The company wanted to get closer to its customers when issues arose, reduce expenses, and increase their share of the market. Sales were $3.4 million in fiscal year 1966 and about to reach $5 million the following year. Once the Picker agreement was terminated, Medtronic intended to sign agreements with distributors in the various countries of Europe, many of whom had been with Picker.

But company executives were not sure which distributors were good to work with, or which would pay their bills on time. Tom Holloran, then a board member and legal counsel, says that the company realized the person most knowledgeable about the various distributors was Manny Villafana. "He was the accounts receivable clerk we wanted," said Holloran. So, in early 1967, Manny was hired by Medtronic as an "international sales administrator." He accepted the position "even without discussing salary," he recounted later.

Soon he was looking for new opportunities. "What about Latin America?" Manny said to his superiors when they were discussing various markets, and they suggested he study it. His investigation earned him the opportunity to create an entirely new market for the company's products just two years after joining the firm.

"We had two people in the company who could speak Spanish," said Holloran, who by then had been named president, "a

Catholic sister and Manny...with great brilliance we picked Manny." By then married and the father of two children, Manny moved his family to Buenos Aires, Argentina, and, as he later said, "learned a lot about sales and marketing." But he was confident that he knew his product well and could teach the doctors about pacing. Unable to get a college degree, he had educated himself; now he was prepared to educate others. As for marketing technique, he recalled what his mother once taught him. "Ask," she said. "Sometimes you get slapped, sometimes you get kissed."

Manny was not intimidated by doctors. "I became their friends," he said. "I would put on white gloves and go into the operating room." This gave him an advantage, since other salesmen were reluctant to get that closely involved. Sometimes, a doctor would have trouble placing an electrode into a patient's heart, and Manny would say, "Move over, I'll show you how to do it," and then make the necessary placement. The doctors came to rely on him so much that he eventually had to tell them he couldn't be there for every implantation and that they were qualified to do it themselves. "But when they had trouble, we would support them," he said, and they appreciated it.

Two government coups in Argentina couldn't get Manny to leave the country, but his son's allergies did. After two near-fatal attacks, Manny and his family moved back to the United States. Upon his return, Medtronic was unable to find him a suitable job. He acknowledged later that it may have been because he was headstrong and hard to manage. In Argentina he'd been on his own. He liked the freedom. He liked the self reliance. "So, I was fired," he says, with the smile of a man who knows he's using a strong word. But to maintain good relations, he gave his boss, Charlie Cuddihy, a wristwatch engraved *numero uno*, smiling again at the irony of such a gift. He liked Cuddihy, and suspected they might be working together again someday. "You learn from every situation," Manny said. Tom Holloran later recounted: "If hiring Manny was one of my brilliant moves, letting him get away was one of my greatest mistakes."

Manny was then offered a sales job at Med General, a small manufacturer of surgical lights struggling to make a profit. By that time he'd been inside many operating rooms, and his personal experience told him the product line had only a limited market. No, he told them, he wasn't interested in becoming Med General's sales manager, but he *was* willing to become president and turn things around. They accepted his offer. Manny was able to raise much needed cash for the company—his first time at that activity—and, by his own account, bring the company to profitability.

Meanwhile, back in Buffalo, New York, Bill Greatbatch had developed a lithium battery for pacemakers. Lithium held the possibility of much longer life—up to three times the duration of existing mercury batteries—and since it produced no gases, it could be hermetically sealed. Greatbatch gave some early development units to Medtronic, suggesting that the company bring them to market. Using normal business prudence, Medtronic tested the units. "They all failed," said Holloran. "And they failed quickly. We had misgivings about lithium batteries." So Medtronic turned Greatbatch down. Learning about this, Manny suggested that Med General capitalize on the opportunity and get into the lithium battery pacemaker business, a potentially big new market. His board said no; if Medtronic wasn't interested, they too were not interested.

Manny disagreed. He resigned his job and called Bill Greatbatch to see if the battery was still available. "Yes," the inventor said, "Medtronic turned it down."

"Why?"

"Insufficient ampere-hours."

"What do you think?"

"I think we can do it," Greatbatch said, and suggested that, once improved, the battery could last ten years.

"Well, I trust you, Bill," said Manny.

Greatbatch later recalled that he warned Manny "that lithium was too new to risk his company on...and told him to work in mercury and then move into lithium." But Manny countered, "No. If I'm going to buck the big boys, it's got to be with something completely new, that solves all the problems but is too risky for them to undertake."

Manny Villafana then visited Tom Holloran at Medtronic and told him he was going to start a pacemaker company using the new Greatbatch battery. He proposed that they work together. The bold offer didn't bring the figurative kiss his mother had suggested, but instead met with rejection. "I remember he escorted me right out the door," said Manny.

"I was out on my own. By this time I had left Med General and had no job. Then both my wife and son went into the hospital. I had to re-mortgage my house. I think I had to re-mortgage my car. I was pinned against the wall." He tried to recruit various people to join him and they turned him down. Not knowing quite what to do next, he remembers going to church one day and finding a prayer card to St. Jude, the patron saint of hopeless causes. He started a novena in honor of St. Jude, and "shortly thereafter things started to happen." He was able to get three men interested in putting a company together, that is, "if he was able to raise the money." Joining him were Tony Adducci, a product manager at Medtronic, who insisted on bringing along his colleague, Jim Baustert, a communications/advertising man, and Art Schwalm, a former Medtronic manufacturing manager who was now building mobile homes in Marshall, Minnesota. That was his initial team. Manny knew of their capabilities and trusted them. The new company was named Cardiac Pacemakers. Early officers in the company were able to buy stock at an average price of 59 cents a share.

Manny called on whomever he knew, trying to sell additional

stock in his new company. A friend offered to introduce him to the Craig-Hallum brokerage firm. Upon hearing his plan, the firm said, "If you can raise $50,000, we'll raise another $450,000 in a public offering." *Fair enough*, Manny thought, and set about asking his friends and acquaintances, and, in turn, their friends, each for a $2,000 personal investment at $2 a share. That meant he had to come up with approximately 25 investors. He said later in an interview that he was able to raise the money because of the earlier success of Medtronic. By then, "It had made a lot of people wealthy, and people were willing to give me a try." Three months later, on February 4, 1972, he opened the doors to Cardiac Pacemakers Inc. in Roseville, a suburb north of St. Paul, in space chosen partly because the landlord told him to "pay me when you can."

Once Cardiac Pacemakers was established, Med General sued the new company, saying it should have some ownership. Medtronic also sued, trying to find out if any trade secrets had been taken. Jack Robinson, president of Craig-Hallum, the underwriter, reportedly said, "I *thought* you had something. Now that Medtronic has sued you, I *know* you have something."

In May 1972, Craig-Hallum offered 100,000 shares in a public offering at $4.50 a share. Trading closed that first day at $9 bid, $10 offer, even though the market for most local stocks was weak at the time. As Manny later said, the company was "nothing but an idea," but with a total of 210,000 shares outstanding, that idea at $10 a share was worth $2.1 million in market value. Ominously, the cover of the prospectus said that part of the proceeds of the offering would be used "for attorney's fees for litigation in which the company is involved." Manny, undaunted at the age of thirty-two, was in business.

Greatbatch supplied the batteries, and two Cardiac Pacemaker engineers, Richard Kramp and Jon Anderson, used them to develop a working lab model and receive a patent on the

(from left to right) Professor Nazih Zuhdi, Villafana, Dr. C. Walton Lillehei, and Dr. Christiaan Barnard. Dr. Barnard did the first heart transplant.

world's first lithium pacemaker. Both engineers had been hired out of the computer industry, *deliberately*. Manny felt they would come to work with an open mind, not the preconceived notions of medical engineers.

Dr. Richard Lillehei, brother to Dr. Walton Lillehei, did the animal lab tests for the new product. One day he closed the door to his office and asked Villafana, "Do you really think we can do this?"

"Yeah, I really think we can."

"Great. I want to beat my brother."

The company had its first human implant in November 1972. The pacemaker was guaranteed by the company to last three years. The device was also smaller and lighter than the competition. Soon thereafter, the guarantee was raised to six years. With such performance superiority, demand swelled all over the world. Revenues exploded from $17,000 in 1972 to $4.6 million in 1974 and $21.4 million in 1976. Earnings followed apace. The stock split 2 for 1 in 1973, and 2 for 1 again the following year. But even such outstanding results could not break through the stifling gloom hanging over the market in 1974, and the stock dropped from 26 to 9 in that year. It was a temporary retreat, however. Two years later the stock hit 68.

The initial lawsuits had been settled along the way. With the company's momentum well established, Manny felt he had accomplished his mission and in 1975 resigned "to pursue outside interests" but would continue as board chairman. Co-founder Art Schwalm was elected president. "The company got to a size where you have a lot of paperwork, a lot of disciplines that you have to learn that I didn't want to learn," Manny explained, and said that he missed working "with doctors in operating rooms, out in the field, which I love." The following year the stock split 2 for 1, again, meaning early shareholders now had 8 shares for each initial share.

In 1977, the company reported sales of $32 million and earnings of $1.77 per share. In spite of the repeated splits, there were still only 2.7 million shares outstanding, and the stock traded around $20 per share, a relatively low multiple of 11 times earnings. Apparently investors did not think the exceptional profitability could continue indefinitely, as Medtronic had now come out with its own lithium pacemaker. Nevertheless, the company was on track to increase sales by more than 60 percent in 1978, and earnings per share by 25 to 30 percent. In the fall of that year, the company received a buyout offer from pharmaceutical giant Eli Lilly. They would give .85 shares of Lilly stock for each share of Cardiac Pacemakers. The offer was valued at $127 million. Based on the number of shares outstanding, Cardiac shareholders would receive new stock worth approximately $45 a share, a premium more than double the current market price. Those holding shares from the initial public offering just six years earlier, *and still holding*, made eighty times their investment.

Greatbatch, the man who had warned Manny to proceed with caution, later wrote: "Manny had gambled and won." Further, it could be added, Manny's bold initiative had changed the industry and improved the health of many across the world.

But the quick success was not so easy coming. Years later, Manny said in an interview: "Many times I felt like giving up. If it hadn't been for friends who took the time to talk to me and to listen, I might have. We were having problems with the product, we were being sued, we were low on money. The competitors were saying ... that our batteries would fail." But in a later interview with the Minnesota Historical Society, Villafana explained how he was able to accomplish what he did. "The entrepreneur has to believe that, within himself, he or she is a super person. If you don't have that spirit within you, there are just too many roadblocks, just too many things that won't allow you to get things done. If you feel you're the average Joe, the average Jane, it's not going to happen." And he added, "after a team is put together," the first big challenge is "raising capital." And raising capital in the public market had been made much easier for him because of the earlier success of Control Data and Medtronic.

Now retired from the presidency of Cardiac Pacemakers, Manny was free to pursue other opportunities. With his history of success, pulling together a new team would be relatively easy and raising capital a near cinch. But there was more to be gained than simple business success. "One of the great pleasures in life is doing things that people say cannot be done," he said. Eager to start anew, he began to search for another venture.

Back at the University of Minnesota, Dr. Demetre Nicoloff had been doing surgical work and related research. One day an engineer came into the operating area and said to the head nurse, "I'd like to talk to a heart surgeon about a heart valve." Such a request was somewhat out of the ordinary, and she was temporarily taken aback. "I have some ideas for a heart valve," he said.

The nurse then asked Dr. Nicoloff, "Can you meet this person?"

"Sure," he said, and they had a conversation.

"Have you seen any valves, or designed any valves?" Nicoloff asked.

"No."

"What makes you think you can design a valve?"

"I'm a good engineer."

Dr. Nicoloff learned that the visitor was Zinon (Chris) Possis, founder of Possis Machine, the maker of automation equipment whose stock had a sensational run in the 1960s. Nicoloff, of Macedonian descent, learned that Chris Possis was of Greek descent, and that created a conversational connection. To help get started, Nicoloff gave his new acquaintance a book about heart valves.

Two or three weeks later, Possis returned and said, "There are some terrible designs in here, just terrible. I think I can design a better valve, but I need your medical input."

"OK," said Nicoloff, and they began a collaboration.

In recollection, Nicoloff thought this happened about mid-1972. To raise money for the development, Possis formed a little company with funds put up by friends. The two men then worked on the valve for "about three or four years, designing it and testing it."

Manny Villafana, looking for an opportunity, learned of the valve and offered to buy the patent rights and also pay a royalty. He asked both Possis and Nicoloff to be on the advisory board of a company he would form, and they discussed how such an arrangement might work.

"But the product needs a name," Villafana said. Valves at the time were usually named after the engineer and the surgeon who developed them. That would make the product the

Possis-Nicoloff valve, or the Nicoloff-Possis valve, but Nicoloff had earlier vetoed the idea, saying that because he was working at the university he didn't want his name on anything. He didn't want to cause problems for university funding. Possis wasn't sure he agreed with Nicoloff—early versions of the valve had been informally called the Possis valve—and there was no quick resolution of the issue. So Manny asked if he could name the valve. There was a pause, and a look of anticipation.

"St. Jude," he said.

"Why would you want to name a valve St. Jude?" Nicoloff asked.

Manny said that his son, Jude, had undergone ENT (ear-nose-throat) surgery at the Mayo Clinic. The family had prayed to St. Jude, the patron saint of hopeless causes, and Manny had promised in prayer, "If you do this, I will do something in your honor." When the operation succeeded as planned, he felt obligated to deliver on his promise. The valve would honor the name of St. Jude, he said, and maybe help others.

The advisory board then agreed that the name would be appropriate. "So that's how it got its name," said Nicoloff. And Manny's new company was named after the valve.

Organized on May 12, 1976, St. Jude Medical bought the rights from Possis Corporation and for the next year tested the valve in vitro (out of the body) for performance characteristics. The company subleased space in St. Paul from Manny's earlier company, Cardiac Pacemakers. In February 1977, stock was sold in an initial public offering by Craig-Hallum at $3.50 a share, again before any product had been brought to market. New SEC regulatory guidelines required disclosure about adverse possibilities, and the cover of the prospectus stated: "In the event the company's development efforts are not successful, the investment of its shareholders may be lost." It is doubtful anyone paid much heed, for Villafana was now an accomplished entrepreneur. He held 27.5 percent of the stock, purchased at .134 cents per share for a total of $25,000.

The prospectus was thirty-four pages long, still readable, but growing in length because of required information about risk factors, the proposed product and how it would work in the human body, business operations, government regulation, and financial statements with applicable footnotes. Salaries were increasing as well, reflecting the inflation of the times. The three officers, Peter Gombrich and William Palmquist, both previously with Medtronic, and Manny Villafana would each be paid $40,000 a year.

Animal implants were expected to begin in the third quarter of 1977 and clinical evaluations in humans later that year. Final clearance from the U.S. Food and Drug Administration (FDA) was not expected until 1979. The company forecast that its heart valve would sell for approximately $1,500, considerably higher than the competition because of its superior characteristics. Anticipating success, the stock traded as high as $9.50 a share in the first nine months after the offering. In December, 1977 there was another offering at $6.50 a share.

When the St. Jude valve was finally ready for market, it was the first all-carbon, bi-leaflet valve, and the company had data from tests indicating it would last two hundred years; that is, it wouldn't wear out. And when a competitor's product was pulled off the market because of breakage problems that could cause certain death, St. Jude was, to choose Nicoloff's words, "in the driver's seat." In 1979, sales hit $8 million and the company was profitable. A year later, sales hit $12 million. Success kept building, and over the years the company had the following stock splits:

1979 – 2 for 1	1990 – 2 for 1
1980 – 2 for 1	1995 – 3 for 2
1986 – 2 for 1	2002 – 2 for 1
1989 – 2 for 1	2004 – 2 for 1

St. Jude Medical was listed for trading on the New York Stock Exchange in 1996 and went on to become a large national company. Investors who held their stock throughout would magnify their holdings enormously. Stockbroker Frank Hurley says he sold a customer 300 shares on the initial offering for an investment of $1,050. Accounting for splits and a current market price of around $40, the customer, who never sold, had stock worth $2.1 million. That's an increase in value 2,000 times over the initial investment, better than the initial Control Data run of 783 times, but Control Data's run covered a span of eleven years, while the St. Jude advance took thirty-five. St. Jude, however, still trades publicly, so the counting isn't over. For Control Data (later Ceridian), it ended in 2007.

With his second success, Manny Villafana acquired a reputation, like Midas, of being able to create gold. He seemed to possess a special entrepreneurial ability, and wanted to concentrate on that. He did not have Earl Bakken's desire to become an accomplished executive. Once St. Jude was properly established, he became restless again. "I have a theory," he said. "A president should step aside after five years. A company that is going to succeed needs fresh ideas and different skills." So he did step aside, and for a year and a half looked about for various new opportunities.

In 1983, he and James Grabek, a colleague from the St. Jude sales organization, formed GV Medical, the initials standing for Grabek and Villafana. They would develop laser angioplasty to clear coronary blood vessels, thus reducing the need for open heart surgery. Craig-Hallum almost immediately raised $3 million at $3 a share in what was then the largest private placement of its history; an effort made easy by the name Villafana, which had become an investor magnet. A year later Craig-Hallum did a public offering, raising $3.5 million at $5 a share, and a year after that made another offering at $8.75 a share, and, in the following year, made still another offering at $10.50 a share. The company was unprofitable throughout those years, but the stock moved up anyway. In April 1987, Villafana announced he was leaving the company "a little early" since it was not yet solidly profitable like his earlier companies had been. The stock dropped from $15 to $12 on the news. But in the annual report of 1987, the company stated it was on the threshold of receiving approval from the FDA for full-scale marketing, and profitability could be expected in the following year. As predicted, FDA approval was received the next year, but only for use on peripheral vessels. The company achieved sales of only $3.7 million and lost $3.6 million. The next year, sales were up slightly but the loss was $4.8 million, and in the subsequent year, sales dropped and the loss increased to $7.9 million. This was not a repeat of the earlier Villafana successes. Investors who sold on the upswing made money, but those who persevered through the losses did not. The company eventually was renamed Spectrascience, moved to San Diego, and today trades in the range of 10 cents a share.

Meanwhile, Villafana had again moved on to another opportunity, Helix BioCore, intending to engage in contract manufacturing of cell growth for major pharmaceutical firms. The company had a public offering in June 1990 at $3.50 a share and subsequently traded as high as $11 a share. Villafana won the National Master Entrepreneur of the Year Award that year, and like a former batting champion, was expected to repeat his successes. The less than satisfactory result with GV Medical was considered an aberration. But in 1991, employee layoffs became necessary at the new company, and Manny was quoted as saying, "I'm sweating." Still, investors wanted to believe in him. A Dain Bosworth analyst was quoted in the *Minneapolis Tribune*: "[Villafana] was very close to failure a couple of times at St. Jude but early investors in St. Jude have made back their money 290 times over."

Villafana (left) and colleagues examine a valve prototype.

Helix Biocore later raised $5 million to develop an artificial heart valve, an entirely new product direction, and in 1992 sold its contract manufacturing assets to concentrate completely on heart valves. Meanwhile, Villafana had started still another company, this one called ATS Medical (meaning Advance the Standard). The two companies combined their businesses under the new name of ATS Medical. Eighteen years later, the company was sold to Medtronic. It was not a bust, but it was not a repeat of earlier successes.

Once again, however, Villafana had left early, and in 1999, formed a new company: CABG Medical (meaning Coronary Artery Bypass Graft). Capitalizing on his reputation, he issued 5.8 million shares to himself at less than a penny a share as a company founder. Investors did not seem to mind, however, as the company in December 2004 sold 5.5 million shares at $5.50 a share, raising $30 million in a public offering. On the first day of trading, the stock closed above $6, even though the company's product had not yet been proven effective. Clinical trials were being conducted in Europe and Australia. United States FDA approval, which was more difficult to obtain, was expected in about a year. A comparison of the dollars invested by various parties illustrates how powerful Manny's reputation had become. He had invested $48,000 in the company at its initial stage; the public had now invested $30 million for fewer shares. Obviously, a repeat of his earlier successes was very much expected.

Eight months later, Villafana made an announcement. "Our device is not working," he said. "The results are disappointing." Grafts implanted in two Australian patients had become blocked, but fortunately the patients were not severely affected. The company reported no revenue and an annualized rate of loss of $5 million. The stock fell more than a dollar on the news, and closed at $2.89, down 47 percent from its initial offering price the previous December.

In a later discussion, the company reported that grafts implanted in twelve of thirty-five patients had a similar problem and two deaths had been reported, although the deaths weren't directly related to the graft. Nevertheless, should the company proceed, Villafana said, redesign would be "extensive and costly, with no assurance that it would be successful," and if successful, the expense of new trials would have to be undertaken. He reported that the company had dismissed seven of its eleven full-time employees and was seeking a buyer for the business. "I am deeply saddened by this event," he said, "and regret that we will not be able to fulfill our commitment to our investors, our surgeons, and most importantly, the patients we aspired to treat."

Only one bidder showed serious interest in buying the company, but the price it offered was less than the cash remaining in the company—hardly a compelling choice. The company rejected the offer, and recommended instead that it cease business and return all remaining cash to the shareholders. This was a frank admission of failure, and a responsible, straightforward proposal. The temptation by many companies to go off in a new direction in order to salvage their reputation was resisted. The remaining money, of course, belonged to shareholders, many of whom had invested at $5.50 a share. On April 27, 2006, the liquidation plan was approved by 98 percent of the shares. Existing shareholders would receive $1.47 per share, and they would be free to invest in a new venture of their own choosing. Of course, one of those shareholders was Manny Villafana, a holder of 5.8 million shares. The *Minneapolis Star Tribune* reported the next day that Villafana would receive $8.5 million for his share in the company. He had invested $48,000. For those who did the arithmetic, this was an increase of 177 times his initial investment, a truly exceptional return over seven years for a company that failed in its mission. There appeared to be little investor dissent, but the newspaper reported that an unnamed shareholder "grumbled about 'the big payout.'" Villafana responded that he had declined "severance and bonus of about $1.3 million." Apparently no one asked why there would be severance and a bonus from a company going out of business. "I put my heart and soul into this company," said Villafana. We could have "run it into the ground, [but] we didn't do that."

Ron Thomas was one of those silent CABG shareholders. He knew about executive responsibilities, since he had started and built a local company called Ciprico and left it in good financial shape upon his retirement. In his opinion, the CABG payoff to Villafana was "not the proper thing to do." But it wasn't illegal.

A year later, Manny began another company, Kips Bay Medical, named, of course, after the boys club of his youth. It would develop saphenous vein support technology for coronary artery bypass surgery. He and a partner invested $10 million to begin early development and tests. Clinical trials would be conducted on 100 to 150 patients in overseas countries, with United States FDA approval expected later. "This is the biggest thing I've ever done," he said, "bigger than pacemakers. We'll create a new industry." In February 2011, he raised $16.5 million in a public offering of 2.1 million shares at $8 a share. The stock closed on the first day of trading at $7.93, a weak opening. This time he had gone to New York to find a brokerage firm to bring the company public, even though he still had admirers in the Twin Cities. "He's Cardiac Kahuna," said Dan Carr, CEO of a Minneapolis group that assists entrepreneurs. "His place in what is still the dominant industry in Minnesota—medical devices—is a rarified one."

But bringing a new product to market had become much more difficult in the medical industry. Since 1976, the FDA had regulatory oversight for medical devices sold in the United States. The agency's requirements for testing before receiving market approval were very exacting and expensive to conduct. Also, clinics and hospitals had now consolidated to become huge medical organizations able to demand price concessions. Medicare reimbursement was another major issue. At the same time, many new competitors had entered the industry, chasing the same dollars. A product innovation no longer brought the quick success of earlier years.

The brokerage industry had also changed. Rodman & Renshaw, the New York firm that underwrote the Kips Bay public offering, withdrew from the brokerage business in September 2012. It was not alone. In the five years 2007 to 2012 (through September), 635 firms left the business. The economic crisis at that time caused many firms to suffer a drop in trading volume, and much of the activity that did take place was executed by low-cost, computer-driven programs. Stockbrokers renamed themselves financial planners and began focusing on asset allocation. With no investment banker, Kips Bay had lost an outside authority to tell its story. Meanwhile, the company was unable to generate revenue and reported continuing losses. There were already 16 million shares outstanding. The stock traded as low as $1 per share in 2011.

10

Cray Research

In the spring of 1972, while Manny Villafana was calling on friends to invest in his new medical company—still just an idea—and Craig-Hallum was preparing an initial public offering for it, Seymour Cray started his own company, also just an idea, and raised the money himself, seemingly without effort. "I was sure about being able to finance my own work," Seymour said. "I would end up having more resources alone than in a big corporation. And that turned out to be the case." He was forty-six years old and wanted to control his own destiny. He would follow his intuition and start with "a clean sheet of paper."

Seymour had become wealthy on Control Data stock and would invest his own money; he knew there were others eager to invest as well. Frank Mullaney, a Control Data founder, invested and agreed to become chairman. Noel Stone and George Hanson, former vice presidents from Control Data, also invested and became members of the board of directors. Fairchild Camera and Instrument Corp., one of the leading manufacturers of memory circuits, agreed to supply components for payment in cash and stock, and also received a seat on the board. Control Data Corporation invested $250,000, apparently reasoning that if it could no longer benefit from Seymour's work directly, it could benefit indirectly. It was a smart decision, very practical, and Seymour welcomed the investment. Quite easily, the company raised $5.9 million in common stock at prices averaging $8.30 a share. Six to eight other employees left Control Data and joined the company, including electrical engineer Les Davis, one of Seymour's long-time associates, and Dean Roush, a trusted

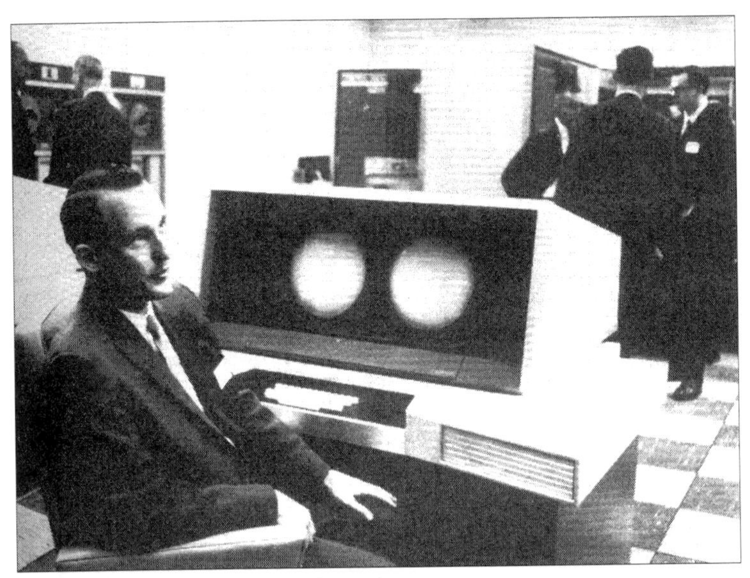

Seymour Cray at the console of a 6600 computer

mechanical engineer who worked on cooling systems. Davis and Roush both took a 25 percent cut in pay. The company tried to be careful in limiting the number of Control Data employees hired at the beginning, but eventually about twelve came aboard. Operations began in space rented in downtown Chippewa Falls, but soon thereafter Seymour had a 12,500-square-foot building erected on his property along the Chippewa River, a short walk from his former Control Data plant. Corporate headquarters was established in the St. Paul suburb of Mendota Heights. Seymour intended to be there no more than once or twice a month.

Of course, the new company would proceed in the manner Seymour chose, building the world's most powerful computer, priced in the $8 million range, and meant to handle the complex mathematical problems of the world's scientific community—a market of perhaps a hundred potential customers. A machine with more features and a lower price would appeal to a larger market, but would also encounter more competition. He had seen it tried at Control Data. As in most of his work, Seymour preferred to keep things simple. He was convinced the company would be successful, but acknowledged that growth might be limited to the target customer base. He would need the same number of employees initially—thirty-five—that had worked with him at Control Data.

Meanwhile, George Hanson, a member of the Cray Research board of directors, was looking for talent to help with the firm's many administrative responsibilities. One day he saw John Rollwagen sitting in a barber chair. He had been Rollwagen's scoutmaster, and the two exchanged hellos. Hanson, as vice president of sales and marketing at Control Data, had earlier helped Rollwagen get a summer job while he was pursuing an undergraduate degree at M.I.T. and an MBA at Harvard. Upon graduation, Rollwagen also worked briefly as a salesman for Control Data. Hanson now suggested that Rollwagen consider coming to work for Cray Research. Draped in a white sheet, with hair falling across his face, Rollwagen agreed to keep in touch. Three years later, a date was set for him and Seymour to have lunch. It went beautifully. They talked enthusiastically about a range of subjects, none specific to business, and when it was over, Seymour said, "Think about it."

"Think about what?" said Rollwagen.

"I think you should join up," said Seymour.

Rollwagen wondered if he had heard right. There was no discussion of salary, no job responsibilities, no anything. But after rethinking it over a weekend, he called Seymour on Monday to say he was willing to "join up." Seymour mentioned a salary and a stock option amount, and then asked, "Can you begin today?"

"Well, no. I have to make arrangements with my current employer and so on. I can be there in two weeks."

"OK," said Seymour. "You will be our vice president of finance. There isn't much to worry about. We have a $5 million line of credit with the First National Bank of St. Paul." Rollwagen had majored in engineering as an undergraduate, but Seymour apparently concluded that an MBA degree, plus clear thinking, was sufficient qualification for a financial job. With such an embrace, Rollwagen felt qualified as well.

But when he arrived for work, Seymour told him the company didn't have a line of credit after all. The bank had turned it down, and the company's cash might last another forty-five days. "But you can sell some of our debentures," Seymour said.

Rollwagen did as told, and in 1975 the company raised $2.7 million in 8 percent debentures, convertible into common stock at $13.33 per share. Although this happened in the lengthening shadow of the stock market bottom of 1974, investors trusted Seymour. A computer had been built and powered up, but was not yet a proven product. It looked like a piece of furniture, circular in shape, with cushions on the lower outer perimeter, which was about chair height. It was approximately eight feet in diameter. Inside the perimeter stood a circular cabinet, six feet tall, shaped like a C, open in the center so the interior could be accessed for service. The central core was covered in dark glass that breathed a comforting hue and blurred the interior of some 200,000 integrated circuits, 3,400 printed circuit boards, and more than 60 miles of wire cut in pieces no more than three-and-a-half feet long. The circular design was intended to minimize the distance for electrical signals and therefore increase the machine's speed. Maybe it wasn't simple after all, but the product was pleasing in appearance, and it was compact. Seymour, in an interview, acknowledged that he enjoyed adding aesthetic appeal to his creations; he said it was an extension of the designer's personality. Most other computers at the time were shaped like boxes. Someone called the new machine the world's most expensive love seat.

Following the sale of debentures, Rollwagen arranged a $1 million line of credit with Northwestern National Bank of

St. Paul—an amount much smaller than the $5 million Cray had mistakenly assumed the First National Bank of St. Paul had already approved. A year later, when he needed more cash, Seymour decided to sidestep the banking industry entirely and make a public stock offering. "We'll do it ourselves. Just like Control Data," he told Rollwagen. Seymour thought a *picture* of the machine under development would help people decide to buy the stock, but upon reviewing the offering document, the SEC demanded further description. Rollwagen doubted the company would be successful selling stock by itself, and was able to get Unterberg Towbin, a firm of national reputation, to do an underwriting. The company had no revenue and the market for new stock issues had not much recovered from the 1974 collapse.

To entice investors, Rollwagen and Seymour went on "road shows" across the country making presentations to institutions and to participating brokerage firms. The two soon grew bored repeating the same points day after day. One day Rollwagen was surprised to hear Seymour discussing the points he normally covered and realized that when his turn came he would have to present Seymour's points. He managed fine. But when it was over, he discovered that Seymour was simply playing a little game, testing the improvisational skills of both and adding a level of entertainment to the work.

The underwriting turned out to be very successful. In March of 1976, the company sold 660,000 shares of common stock at $16.50 a share to the public and raised $9.8 million, net of expenses. "We learned that those who wished they had bought Control Data, but hadn't, were eager to subscribe," said Rollwagen.

This was an ordinary investor's first chance to participate, and a surge of orders caused the stock to trade as high as 25 on opening day, but it drifted lower in the weeks that followed.

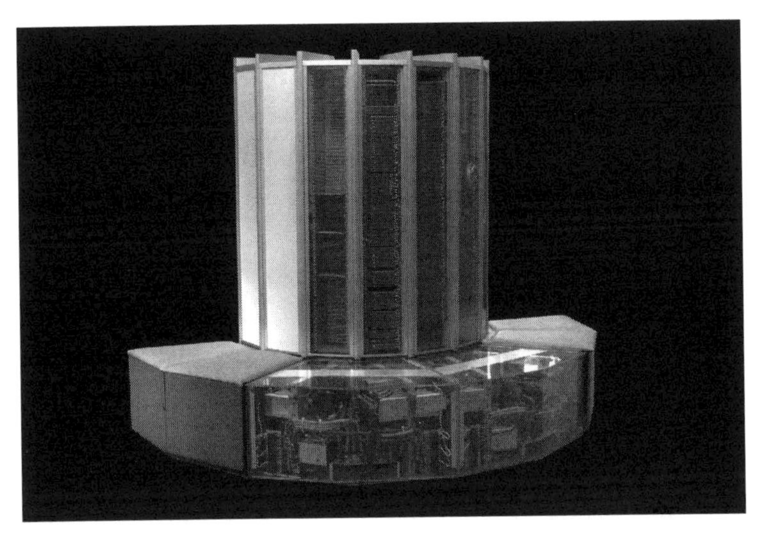

The Cray-1

Prior to the offering, there were only 707,006 shares outstanding and 50,800 shares under option. Seymour, of course, knew the arithmetic of limiting the number of shares in order to maximize their potential. There was no revenue from operations, but an initial computer named the CRAY-1 had been shipped to Los Alamos Scientific Laboratories and was due to undergo a six-month evaluation. It was the first time a company had shipped computer equipment for a no-charge evaluation. Investors, gambling on another extraordinary success from the company's namesake, anxiously awaited the lab's conclusion.

In September, the Los Alamos team issued its report: "The conclusion of this evaluation is that the CRAY-1 meets every performance criterion." The lab signed an agreement to lease the machine. The stock recovered to a high of 22 bid in the third quarter, but traded to a low of 13.75 in the fourth quarter. In his year-end annual report, only fourteen pages long, Seymour recited the company's great achievement with measured restraint. "In 1976, Cray Research reached its initial goal. A large scientific computer was designed and built by a small company in four years, delivered to a major computing facility, and judged technically superior to other currently available equipment." It was a significant statement delivered with understated pride, yet the stock did not respond. Perhaps investors had already anticipated astonishing performance from their president. What Seymour didn't mention, however, was that the entire cost of development would be covered by the sale of just one machine, an extraordinarily quick recovery of early at-risk expense. Such a conclusion was apparent from the general numbers available: $2.9 million had been spent on hardware and software development to date, and the product, with an $8 million sales price, would certainly carry better than a 50 percent gross profit margin. That informal assessment foretold that the delivery of future machines could bring extraordinary profitability. Nevertheless, because the first delivery was on lease—meaning revenues were delayed—the company was unable to report a profit. Investors remained cautious.

Following that initial accomplishment, Seymour said that the company would move on to its next goal, which was to become a "leading supplier of computers to the segment of the market known as 'large scale scientific,'" and that more money would be spent on programming to make the equipment more effective. Anticipating that additional customers might choose to lease their machines, the company arranged a $10 million line of credit.

For the first three quarters of the following year, the company lost money because lease revenue was insufficient to cover expenses. But in the fourth quarter, a computer system was sold to the National Center for Atmospheric Research in Boulder, Colorado, for $8.8 million (versus a lease price of $210,000 a month), and the company earned $2.07 a share in that quarter alone. In total, three systems were shipped during the year, two on lease. Counting all four quarters, the company had revenues of $11.4 million and earned $.71 per share fully taxed (although in reality no taxes were paid because of a tax loss carry-forward). The stock traded between $15 and $30 a share. In other words, at its high, it was selling at 42 times earnings. But it was clear that if the company could sell, not lease, four systems a year it could earn more than $8 a share. If that happened, the stock price at $30 was selling under 4 times earnings, an unbelievably cheap valuation. Such an assumption was based on a big IF, but not an impossibility. The company was planning to build and ship four machines in the coming year. In his year-end report, Seymour said, "Our goal is to be a very profitable computer company." He also said that John Rollwagen, age thirty-seven, had been promoted to president, leaving Seymour free to direct "in detail" the technical efforts and new product development. Besides finance, Rollwagen had also been involved in marketing. His skills would supplement Seymour's nicely.

Not four but five machines were installed in the following year, 1978, only one of which was an outright sale. But the cumulative build-up of lease revenue was more than triple that of the previous year, total revenues advanced to $17.2 million, and earnings increased to $1.63 a share, better than double the previous year. In spite of the revenue lag associated with leases, net earnings after tax exceeded 20 percent of revenues. It was apparent that the company could become very profitable regardless of the sales/lease mix. The stock reacted accordingly, hitting a high of $77 a share. In November, the company declared a 5 for 2 split, and also announced its intention to double the rate of production to eight machines in the following year.

At the National Center for Atmospheric Research, someone hung a sign on their Cray supercomputer, naming it Deep Thought. It was a complimentary and somewhat amusing tag, but even more so when one realized the subtle reference to a clandestine source of information during the Watergate hearings called Deep Throat (whose true identity remained unknown for over thirty years.) But not everyone was so amused about the machine's work. Seymour received phone calls and threatening letters condemning the role his supercomputers played in nuclear weapons research. He chose not to comment immediately about that, but in August 1979 remarked in an interview, "The ability to test bombs on a computer seems to me to be the vehicle that led to the test ban treaty. As long as we can keep it on a computer, no one will get hurt."

Revenue in 1979 increased approximately two and a half times and profits tripled. The company called it an "extraordinary spurt of growth," but said it was challenging as well as rewarding and caused management to question "how large we should be" and "how fast we should grow." The company now employed 521 people. Rollwagen, the president, took pride in meeting all new employees and had trouble maintaining his acquaintances. To create a stable base of revenue, the company began to encourage customers to lease rather than buy machines. And, after careful consideration, the company decided, once again, to resist developing less powerful products to reach a larger market. It did, however, bring out an upgraded CRAY-1/S series, and also began developing hardware and software to enable its machines to communicate with computers made by other manufacturers, explaining that this would "expand our markets." To help sustain its preeminent technical position, a subsidiary called Cray Laboratories was set up to do research in Boulder, Colorado. "There are elaborate plans to decrease reliance on me," Seymour said in an interview. "I hope they don't move too swiftly." Meanwhile, back in Chippewa Falls, he was working on a newer machine, six times faster than its predecessor. During the year, the stock again tripled.

At about this time, a world-famous nuclear scientist from France wanted to meet Seymour at his place of work to see how he operated, how he handled tough problems, and so forth. The scientist represented a large customer for Cray Research, so Rollwagen made arrangements for him to visit Chippewa Falls.

"I work in my house, by myself," said Seymour, "about three hours or so, and then I'm sometimes stuck on a difficult mix of problems. So I stop and go down the bank toward the lake and work in my tunnel. It's supported by 4x4 cedar beams, and I dig deeper into the bank. While I'm gone, little elves come out of the woods and into my office to solve the problem." His eyes gleamed with delight. He had finished his explanation.

Whether the scientist really understood or not, nobody knows, but he must have thought the story a quaint, almost childlike, explanation from a man of brilliance. "Of course," said Rollwagen upon recalling the event, "Seymour was just adding whimsy to a very natural process. When the mind relaxes, the subconscious works on the material already there and brings forth a conclusion. That's the work of the elves."

In the spring of 1980, at the company's annual meeting, President John Rollwagen said that the company intended to deliver ten systems in the current year and twelve to fourteen in the following year. Such a forecast indicated that demand for the machines remained at a very high level. But at the same meeting, Seymour Cray expressed his worries about the price of the stock. "I would caution stockholders," he said, "to take a hard, serious look at what they are buying." And he followed that remark with an unambiguous warning. "I think the current price is greatly inflated over the value of the company." The stock price at the time was $47 a share, or about 25 times earnings for the year just finished. But running counter to this caution, the company reported first-quarter revenue and earnings more than double the year before.

"We keep wondering whether the company is showing signs of maturity," Seymour said, "but you can't say we've become a mature company because we haven't had a setback yet." His worries seemed to come from the fact there was a two-year slippage in getting parts for the CRAY-2, the follow-on supercomputer currently under development. But Fairchild, the company's supplier, who had a director on the board, expected the shortage to "ease over the next year." Choosing among the various remarks at the meeting, the Minneapolis paper ran a headline the next day:

CRAY FEARS STOCK PRICES TOO HIGH

Nevertheless, in September, Morgan Stanley, the prestigious Wall Street firm, issued a new buy recommendation on the stock, causing it to jump $8 in one day. "I believe the market for these supercomputers is about to burst forth," the analyst said, and he estimated that earnings would advance to $2.40 a share in the current year, up from $1.89 in the previous year, and increase again to $3.40 a share in 1981, which was about 50 cents to a dollar over prevailing estimates. He also forecast earnings of $4.90 a share in 1982. John Carlson, the company's financial vice president, declined to comment on the earnings estimates but said, "We think the outlook is very good. We are right on plan. The key variable is the number of systems purchased versus those leased."

On November 18, 1980, the stock was listed for trading on the New York Stock Exchange, and in December the company split the stock 3 for 1. In spite of Seymour's earlier warning, investors made their own evaluation. The stock almost tripled from its high the previous year. Investors lucky enough to have bought at the initial public offering were now up 22 times on their investment, and they owned stock in a company on its way to become a *Fortune 500* company.

The following year brought continuing success. Reflecting the 3 for 1 stock split, the company earned $1.32 per share, which would have meant earnings of $3.96 per share before the split. This exceeded the Morgan Stanley analyst's estimate of $3.40 per share, which at the time was also higher than anyone else had predicted. Thirteen machines were installed, six of them outright sales. Revenues exceeded $100 million and total employment passed a thousand. The company was exceeding even the most optimistic expectations. More than that, customers were now coming from a broad range of industries as they realized that modeling complex physical phenomena in numerical terms would be cheaper than actual experimentation. Three installations were sold to the oil industry, including one to Exxon valued at $17 million to be used for seismic analysis. Other machines were used for simulating aircraft design, molecular behavior for biomedical research, and car crashes to improve automobile design. One customer used its machine's computational strength to play chess with "the best in the world." The penetration of new markets resulted in sixteen new orders for machines scheduled to be shipped in the following year. This enthusiastic response meant that the original market estimate of approximately a hundred customers in the scientific community was now much enlarged, and the company did not have to bring out a lower priced product and get into debilitating competition with the rest of the industry. Seymour's vision of a profitable, very specialized supercomputer company was succeeding marvelously, probably even beyond his own expectations.

But on November 18, 1981, trading in the stock was halted at mid-day pending a press conference to be held the following day. The stock price had already jumped to $39 that morning, up $3 in anticipation of a "major announcement." It was understood that Seymour Cray himself would be coming out of Chippewa Falls to conduct the press conference; that, in itself, was major news. Wall Street traders, of course, wanted to get a jump on the announcement. Some anticipated that a new product might be announced, or a major financial deal, or a personnel change perhaps involving Seymour. But if Seymour was leaving, some reasoned, the stock would be down, not up. Speculators slept uneasily that night.

The following day the company issued a press announcement covering a range of issues. Seymour had a radical new design for the machine under development, the CRAY-2, and it could now be expected to perform up to twelve times faster than its predecessor, though it would be only one-tenth the size. The longest wire would be sixteen inches, versus forty-two inches in the CRAY-1. Such diminution would be achieved by a three-dimensional circuit-packaging technique, a novel innovation, but the increased circuit density meant, of course, new problems in removing heat. The company had tried various approaches, all of which proved unsatisfactory, but now, in what Seymour called his "last desperate attempt," the company had decided to place the circuitry in a clear inert liquid. This led to an observation by an outsider that if the previous machine was the world's most expensive love seat, this one would be the world's most expensive aquarium. The price for the new system would range from $10 to $20 million. Seymour said the announcement was intended to help potential customers make future plans because, in placing an order, they would be committing to a substantial financial investment. Some observers felt, however, that the announcement might have been intended to offset inroads by Control Data, which had recently won a few competitive bids.

Also announced at the press conference was a stunning personnel change. Seymour would resign as chairman of the board and become an independent contractor. He would remain as a director and as a member of the company's executive committee, and he would devote all his time to development of the new machine. However, he would be free to start another company if Cray Research didn't market the machine to his satisfaction. "But I'd only do that as a last resort," he said, and joked that the

The Cray-2

whole purpose of the arrangement was to "avoid driving to Minneapolis [St. Paul] for a meeting every week."

In an apparent effort to reassure investors that the company was still in capable hands, fourteen employees were promoted to titles of vice president or the equivalent. Investors didn't care much about that; it was Seymour they were concerned about, and the stock traded near $36, down almost $3, after the announcement. Some surmised that the new agreement would mean that Seymour and his unique creativity would not be included if a buyout offer were made for the company. Responding to such speculation, Rollwagen, now chairman, said that the company had earlier received a call asking whether it was true that the French company Schlumberger Ltd. was offering $90 a share for Cray Research. "Sold," he quipped in reply.

Summarizing the effect of the announcement, a computer industry analyst from the San Francisco firm Dean Witter said: "Clearly the fact that they might have a big new supercomputer is being weighed against the fact that Cray is leaving day-to-day management of the company." When asked his opinion, Frank Mullaney, the initial chairman of Cray Research and a long-time colleague of Seymour Cray from earlier ERA and Control Data days, said, "He always has to have his little bomb to drop." The bewilderment that followed in the days after the announcement was enough to cause Seymour to agree to interviews with two Twin Cities newspapers to help clarify his new position.

On December 7, 1981, *St. Paul Pioneer Press* technology reporter Don Clark opened a lengthy article by mentioning that one of Chippewa Falls' native sons, "a near mythological figure," liked to spend his summer days windsurfing on nearby Lake Wissota, and that he was famous for declining interviews and dodging questions about his personal life, but because of "misunderstandings," he was willing to break his silence.

"People just assumed the most obvious thing," Seymour told Clark, "that there was some sort of rift with management. But that's not the case." Under the new arrangement, he would work in a new facility being built for him in Chippewa Falls. Both he and Cray Research would have the rights to his development work. The reporter observed that Seymour appeared at peace with his career and was congenial in conversation. "My interests are to do the 'thing' part with computers, rather than the 'people' part," Cray said. "John Rollwagen ... and I had a very good relationship. Our perspectives...are very similar. My role [now] will be to tell him what I think, and he will make the decision. I [am] really happy to feel less guilt, so to speak, in not holding a position in which I was not performing."

During the interview he also commented on the frustrations of working for a public company and the relentless pressure to grow.

John Rollwagen (left) and Seymour Cray at the new Chippewa Falls plant.

> *I am absolutely convinced [he said] that it is not so much the personalities of the individuals but just our society's view of public corporations. The incentives for growth when you are a company are very hard to deny. I can't say, 'well, let's just stop this research now, I'm comfortable with this size company. Why don't we stop now and not grow anymore? We can just pay dividends.' I can't do that, because it is really inconsistent with what the stockholders are going to do. It has nothing to do with personalities.*

And he further explained, somewhat obliquely: "The way to avoid something is to prepare for it. The agreement says that if the company really does not want to do my thing, then I can do it myself. It wasn't to do it. It was to prevent it from being necessary."

The mission of the new lab at Boulder was "to explore the latest technology," he said, whereas his own view is to "always use ten-year-old technology and take advantage of the unique packaging kinds of things that I'm good at." He explained that the liquid being used to remove the heat from his new circuitry was made by Minnesota Mining and Manufacturing Co. and had been around for twenty years. It had not been used earlier because everybody thought there would be problems. Since nobody had ever done it, "I guess we were all chicken," he said. After re-mixing the ingredients to form a recipe for the company's unique requirements, and incorporating a method to filter out contaminants, he was able to say: "I am now confident we will not have many problems with the cooling system." It was a "low technology approach."

Then he imagined a future no one wanted to consider. "If [the Boulder Lab] can make a competitive machine...then indeed the company will be quite independent of me. The timing will be quite good then. I'll be having health problems in another five or ten years." (He was fifty-six years old.) He acknowledged that he left Control Data when it got too big, but he couldn't imagine that happening at Cray Research. "Not until it's a billion dollar company. So I've got a little time left." Revenue at the time would have had to increase ten-fold to reach a billion.

Three weeks later, the *Minneapolis Star* interviewed Seymour, this time at his cottage on Lake Wissota, about a mile from his home, where he often did his thinking and working with the aid of a small computer. "Despite his reputation as a recluse, Cray is a pleasant man," technology reporter Steve Gross said, and he quoted Seymour at length, as summarized below.

I was just lucky enough to start in this business when there was no prior art. There are not many technical people who are patient enough, and who have the ability to do the whole job. I am a detail man. What I worry about is if the fourteenth decimal point is right. It's doing the details right that makes the difference. I think it's amazing that I can be successful with that approach. I continue to be amazed by it. Perseverance is the key to success. Of course, you also need talent and confidence. An enormous amount of confidence is required. I'm curious to find the point where I peak. In sports, people peak in their twenties. I like to think that at age fifty-six, I'm near my peak.

My computers just add, divide, multiply and subtract, [but] I think there is a speed that comes with simplicity. Not all of the modules work the first time. The people on the [thirty person] development team show me my mistakes. I feel like I'm a very small part of the company.

As for the computer being named after him, he said, "It wasn't my idea. I didn't like it, so I just ignored it."

And then he talked about some differences of opinion. John Rollwagen believed that the company could market both computer systems well into the future, largely because customers would not want to sacrifice the software they had written for the CRAY-1 machine and might not immediately require the power of the new CRAY-2. Seymour, on the other hand, said, "It is possible that Rollwagen is right. But…I think that by 1986 we will have phased the liquid cooling technique into all of our products….It will be quite a challenge to phase one product in and another product out."

There was also a disagreement about delivery dates. Rollwagen said that the new computer would be delivered in 1985. Seymour said it would be delivered in 1983. "I've announced my date and John [Rollwagen] has announced his. If there is a difference, that's fine." The contrasting opinions suggested that the company did not have a fully coordinated approach. Rollwagen was probably trying to manage customer expectations without undermining company revenue. Seymour, on the other hand, seemed to find it difficult to remain silent when his pride was at stake. His personal goals exceeded Rollwagen's.

He also reinforced his point about limiting his management responsibilities. "I can't think about my Boolean [mathematics] if all these people problems are rolling around in my head. So I don't want any people problems."

Recounting events some years later, Rollwagen suggested that the company might have lost Seymour if the changes hadn't been made because Seymour had a habit of leaving corporate entanglements and starting anew. Ironically, he was at a pinnacle of success. He had designed a computer that was in exceptional demand, and had just announced a superior successor. Yet the pressure of unrelenting expectation never lifted. He felt responsible for a whole corporation of people, and while he took pride in astonishing the world with surpassing accomplishments, he seemed unable to truly enjoy his achievement.

Rollwagen sensed his unease.

"Seymour, it's not really working, is it?"

"No, it's not, John."

"You're going to leave, aren't you.

"Yeah, I might."

"Well, Seymour, your name is on the door here. This is not Control Data. Maybe we can work out a way to keep you around."

And they struck up the independent contractor arrangement that caused so much anxiety among investors. What the outside world didn't know was that it could easily have been much worse.

The following year, 1982, brought new surprises. The United States went into a steep recession that made customers reluctant to place orders for new computers. Two shipments intended for France were held up for export licenses. The company then forecast shipments of only fifteen supercomputers for the year—not sixteen as announced earlier. Analysts lowered their earnings estimates accordingly. The stock fell from $45 year earlier to $26 at annual meeting time in May. At that meeting, Seymour called attention to his earlier forecasting prescience and his concern for the financial safety of shareholders. "For three years now I've been coming to annual meetings in Minneapolis and telling shareholders to get out there and sell a little stock. The price is too high. Now the price is more reasonable." It may have been the first time an executive wanted credit for insisting his company's stock was overvalued, and that he was more comfortable with its now-reduced value.

These remarks may have reflected Seymour's unease with the responsibility investors placed on him for ever-improving results, but they may also have been based on a realistic appraisal of the limited market for supercomputers. Los Alamos National Laboratories had already installed five machines, and Lawrence Radiation Laboratories in Livermore, California, had installed six. NASA had two. Perhaps they had all they needed. Seymour also knew that unforseen factors could upset any company's carefully constructed plans. Now that the price of the stock seemed fair again, he would put his concerns to rest.

In April, the company announced its intention to introduce a multiple processor version of its supercomputer, called the CRAY X-MP, two to five times faster than its predecessor, the CRAY-1. The new machine would be based on the basic architecture of the CRAY-1 and would be program compatible. The company also announced a solid-state storage device fifty to a hundred times faster than disc storage devices currently available, priced at roughly $2 to $3 million. There would be a market for this product at the company's forty-some existing customer sites, and it might appeal to new customers as well.

The new CRAY X-MP model had been developed by Steve Chen, a vice president at the company's Chippewa Laboratory, working independently of Seymour. It had been Rollwagen's decision to pursue the project and create an alternate design team. Although raised in Taiwan, Chen had received advanced degrees at U.S. universities, including a PhD from the University of Illinois, a center of computer research. He worked for Burroughs Corporation from 1975 to 1978, where he helped develop a supercomputer called Scientific Processor that was never brought to market. He then left and spent a year at Floating Point Systems before joining Cray Research. Commenting on the new product announcement, one Minneapolis analyst noted that "the computer and the storage device were the first Cray Research products to emerge without the aid of engineer Seymour Cray." The company was no doubt eager to create the impression that its future wasn't entirely dependent on the genius of its founder, but Seymour himself was privately taken aback that someone could make such significant improvement on one of his initial designs. Nevertheless, when the new machine worked, "he was very gracious," said Rollwagen.

That fall the company cut prices on all product offerings, claiming that new technologies had reduced the cost of components. A version of the CRAY-1 was now half the price of the original machine, with no reduction in performance. Rollwagen described the company's intentions: "We have... established a new price/performance curve for the large-scale computer marketplace." It was now clear that the company had decided to broaden the market for the original CRAY-1 architecture, guessing that the price reduction would appeal to more customers and increase overall revenues. Or was it clear?

Calling it a "blockbuster announcement," a Kidder Peabody analyst had this comment: "I'm surprised it's as aggressive as it is." Some analysts speculated that competition from Control Data's Cyber 205 probably influenced the decision. Further, two Japanese firms, Fujitsu Ltd. and Hitachi Ltd., had announced plans to market new supercomputers in the upcoming year, and

IBM and Sperry Univac also had announced top-of-the-line models that might compete.

In his year-end annual report for 1982, chairman and CEO John Rollwagen said: "We are no longer a one-product company. Rather, we offer supercomputers priced from $4 million up to $11 million. Furthermore, with the coming introduction of the CRAY-2, we will be offering two families of computer products during the mid-and late-1980s." It was a straightforward acknowledgement of a successful evolution. Still, both families of computers were based on the initial designs of Seymour Cray, the company's inimitable founder, and he was now an independent contractor. Although revenue for the year increased 39 percent to $141 million, earnings barely improved, coming in at $1.38 per share, versus $1.32 the previous year. The stock traded at a low of $20 that year. Although earnings had stalled, investors should have heeded Seymour's suggestion at the previous annual meeting that the stock was once again fair value.

Revenues for Cray Research increased every year for the next six years and so did earnings per share, but the stock peaked in the fifth year. A Kidder Peabody analyst estimated that in 1985 the company had nearly 80 percent of the supercomputer market. The stock split 2 for 1 in that year, and two years later climbed to a summit of $134, which meant it had increased tenfold in value over the pre-split level when Seymour last expressed his concern. Investors who held the stock since the initial public offering now held fifteen shares for each original share and, factoring for price appreciation, had an increase in value of 122 times. At the same time, the price of Control Data, the company Seymour Cray had left, remained far off its peak. A nimble investor would have been highly rewarded in making the switch, but few are able to jump from one stock to another with perfect timing. It would have been rewarding to simply catch and enjoy a good upward swing.

The number of CRAY systems installed at customer sites reached a hundred in 1985, the total originally envisioned when the company began, but there was still plenty of opportunity ahead, not only in the United States but around the world. There were estimates that the market would grow to five hundred machines by 1990. In spite of this potential, with hindsight we now know that investors should have been satisfied with their reward and cashed in, for in October 1987, there was a near panic in all stocks, and Cray Research stock dropped to $54. The company earned $4.65 a share, up from $3.99 a year earlier, meaning it was selling for just 12 times earnings at its low. The stock worked its way up to $87 in the following year but never reached that level again. In fact, two years later it sold at $20. Most stocks recovered in the years that followed, but Cray Research did not. The company's earnings per share flattened and the expectations of future growth slowly vanished. It wasn't because the market for supercomputers was contracting—that came later. It was because profitability was being squeezed.

With two hardware development efforts to maintain, along with software expenses approaching the level of hardware, total research and development costs were growing rapidly. In the year 1987, although revenues had grown to $687 million and could support a high level of expenses, the company spent $110 million on research and development. When compared with the cumulative total of $2.9 million spent during the years 1972–1975, which produced the company's first machine, it was clear that the competitive race had become much more demanding, and that a huge, reliable flow of cash was necessary to sustain success. Revenue could not be counted on to grow indefinitely, and an earnings squeeze would limit the opportunity of outside financing. The company decided to conduct a review of possible options.

A portion of the development budget again supported Steve Chen, who was designing a revolutionary new machine. But the design would require five additional technological breakthroughs, and the project had already consumed $50 million. Chen had been with the company eight years and had been celebrated as a talented successor to Seymour Cray. He had accomplished much in designing the CRAY-1 follow-on machine, named the X-MP, and a sister product called the Y-MP, which had helped bring the company success in new markets. Nevertheless, taking a top-level view, the company decided that the expense of Chen's current initiative and the gamble it posed was not affordable. The project was cancelled. The stock dropped 13 percent in the two days following the news.

His work now unfunded, Chen and several of his staff decided to leave the company. At the age of forty-three, convinced his goals were still achievable, he began looking for financial support to create a company of his own. To fill his vacancy, the company gave engineer Les Davis responsibility for computer development.

Davis had an interesting history. He was a modest, soft-spoken fellow, trained as an electrical technician, who joined ERA in the early days and took night courses at the University of Minnesota to get his college degree. In 1962, he graduated with a bachelor's degree in business, not engineering. He says his "best learning experience was working with Seymour and the many other talented people at ERA and Control Data." Over time, he became the technical employee Seymour relied on most to develop his ideas. Seymour felt free to conceive design possibilities, while Davis executed them with the rest of the staff and handled the personality issues Seymour so wanted to avoid. To work closely with a man of Seymour's intellectual capacities must have required special abilities, and Seymour seemed to appreciate them in Davis. "We became good friends," Seymour said. "We had a symbiotic relationship." Over the years, Davis had delivered exceptional work. Now he was alone as the head of research in one of the leading technology companies in the world. Seymour, of course, was an independent contractor.

Back in Minneapolis, an entrepreneur who loved challenges heard about Chen's situation and arranged a meeting. He surmised that the two would be a good fit based on their complementary abilities: Chen's technical prowess and the caller's skill at raising money for new businesses.

"I understand the project clearly," said Manny Villafana after meeting with Chen. "It is an exciting one, to say the least." By this time he had already started Cardiac Pacemakers, St. Jude Medical, and GV Medical.

Although the two would make an unusual team, Manny felt especially empowered because of his proven success, and also because he had deliberately hired computer engineers to develop the first Cardiac Pacemaker product. Though he didn't have a formal education on technical matters, he had a good understanding of them, and he knew the merit of blending disparate disciplines for cross-fertilization. After a second meeting, however, Steve Chen decided he preferred to work with someone more experienced in the computer business, and the two ended their conversation. Chen went on to receive prestigious endorsement from an industry colossus still trying to master the top levels of computing power when IBM agreed to finance his work. Villafana moved on to other ventures more closely related to the medical field.

In 1989, earnings for Cray Research dropped almost 40 percent from a year earlier and the company again felt the need to rethink its research expenditures. There were still large development activities at two locations, Chippewa Falls and Boulder, and canceling either would bring personnel issues, as engineers would begin to doubt their security at either location. Rollwagen met with Seymour to discuss the problem. They agreed on the basic issue: the level of technology they were exploring would continue to pressure earnings.

"How much would it take to finish your project?" Rollwagen asked.

"A hundred million," Seymour replied.

The two discussed different possibilities within the existing corporate framework and even some outside it. Finally Rollwagen said, "We will form a separate company." Seymour agreed, but this time said he wanted his name on the new company.

It was a difficult decision, Rollwagen said, but the right thing to do. The new company would be called Cray Computer and would be based in Colorado Springs. The work already being done there would become part of Seymour's own, rather than his in-house competition. But the stock dropped 10 percent on the news.

Ninety percent of Cray Computer's stock was distributed to Cray Research shareholders; the remaining 10 percent stayed with the predecessor company. The new entity was given assets worth $55 million and a loan of $100 million and "would devote itself to developing the long awaited CRAY-3."

"It's a stunning development," said Gary Smaby, an industry analyst. "For a company to fund and set up a direct competitor must be unprecedented." It was indeed, but the company was addressing new realities. Control Data had announced the closing of its supercomputer operation, renamed ETA Systems, with a final write-off of $335 million. Since U.S. government agencies preferred two bidders on every contract, some thought the Cray split-up was a "political maneuver" designed to strengthen America's competitive position against ever threatening products from Japan. The company did not deny it. Japanese government policy had been very nationalistic concerning the purchase of computers, and Cray Research had long felt excluded from their market. The idea of protecting a critical American industry was entirely plausible. Importantly, the two companies, although competitors, would be entering into cross-licensing and technology transfer agreements involving both hardware and software.

"We [had] an embarrassment of technological riches," said Rollwagen, and he explained that the creation of two companies would allow the pursuit of both existing technologies.

Speaking of America's competitive position in the world, he said, "[The arrangement] will help this country maintain its dominance in this vital technology." But he said in a later interview that the split was driven entirely by internal company needs. Development expenses had simply grown too large.

In summary, then, the two companies would compete against each other, but for a time at least would operate under different business conditions. One would have no immediate revenue,and would concentrate on maintaining the tradition of bringing out the industry's most powerful computer targeted primarily at the world scientific community. As a new company it would be expected to lose money during its early years. If the development was promising, additional funds could be expected from investors. The other company would be an established business with revenue from a customer base that included both scientific and general business users. Its profitability would increase because a major development expense had been removed. Both companies would have the benefit of shared technology. The solution was an innovation, unusual in its arrangement, but born of necessity.

The industry was changing in ways that would make it difficult for any traditional player to survive. At the Cray Research annual meeting in 1987, Seymour Cray said he was astonished at the progress of personal computers. "A $10,000 personal computer today has as much memory as the original CRAY-1, and by the 1990s personal computers will have a memory the size of a CRAY-2." He followed with an ominous comparison. "It's a revolution in that industry that's as impressive as ours [supercomputers], or more so." Nevertheless, he continued working on the technological promise of his CRAY-3 machine. It would be so small, he joked later, that it barely had room for the company logo. Now his new company, Cray Computer, would focus entirely on that product.

Cray Research was without its namesake and without Steve Chen, the apparent successor. It was left to struggle in an industry undergoing profound transformation. Mini- and microcomputers (computer on a chip) were undermining the traditional mainframe business, threatening the profitability of companies like Control Data and others that were still competing at mid-range prices. Microcomputers strung together as parallel-processors soon delivered supercomputer power at a fraction of previous prices and undermined profitability at the top end of the industry. Further, the price of big machines was being threatened directly by Japanese computer manufacturers. Nippon Electric Co. reportedly made a sale to the Houston Area Resource Center by giving back "half of the $22 million purchase price in research contracts." And perhaps most important of all, in a few years the Cold War would end. This reduced the appetite of the largest customer in the world for powerful computers—the U.S. Government.

These changes in the industry affected everyone. Hoping to regain the benefit of work from its former employee, Control Data agreed to use Cray Research as a subcontractor for supercomputers when it bid on large government contracts.

Seymour's personal life had undergone transformation also. Once he achieved technical success, he decided to give some attention to the social limitations of his introverted behavior and his habit of intense concentration at work. "He came to see that there was more to life," his sister, Carol, said. He went to parties and even held some of his own, inviting a variety of guests and watching them interact. "I discovered people," he admitted in a rare public comment about something other than his work. In 1975, he and his wife divorced. They had raised three children. Five years later he married Geri Harrand, the owner of a physical therapy business. She was more than ten years younger, and helped create an active life for them both. They went to plays, windsurfed together, and skied with friends. "He's still basically shy," she said, "but he's come out of his shell a little bit."

Professionally, Seymour continued to receive acclaim for his many earlier achievements. On May 29, 1989, *Time* magazine, while discussing the difficulties he was now encountering at work, nevertheless said, "He was to supercomputers, what Alexander Graham Bell was to the telephone," and cited his record of "five major computer designs dating back to 1960, each for what would be the fastest machine of its time." The article was referring to the Control Data 1604, the 6600, and the follow-on 7600, plus the CRAY-1 and the CRAY-2. *BusinessWeek* had his picture on the cover of its April 30, 1990, issue, calling him in very bold letters "THE GENIUS," and said that, although he had guaranteed himself a place in the history books, he was now in trouble because his CRAY-3 project was two years behind schedule and the competition was catching up. He was relying on gallium arsenide chips, much faster than silicon but difficult to produce with consistency because of their fragile nature. This led to enormous development expense and product delay. The praise for the man sounded as if his magical powers were gone—almost like an early funeral oration. Larry Smarr, director of the National Center for Supercomputing Applications at the University of Illinois, was quoted as saying, "I believe history will rank Cray with the likes of Thomas Edison."

Now he was working on the CRAY-3 in the spin-off company based in Colorado Springs. As usual he was pressing the technical limits, but this time with untested technology. In addition to the unpredictability of working with gallium arsenide, he intended to place 1024 chips in a 4x4-inch package. Assembling them would require such microscopic precision that he envisioned robots, not yet available, that would have to be designed from scratch and tested. Also, new instruments would have to be created to test the circuits themselves. Seymour was again pushing himself to accomplish the exceptional, to climb the mountain of technical success that others could not reach, to achieve his peak intellectual performance. He was in his mid-sixties, an age of decline for most. To succeed would require all his powers of concentration and an unremitting resolve, traits that had served him so reliably in the past. He had often described his gifts as intuition. "Don't do anything that other people are doing," he had said once, insisting on the original approach, much like Manny Villafana in his initial pursuit of an unproven pacemaker. But the work Seymour was doing continued to need money. By 1989, $120 million had already been spent on the CRAY-3 project. The stock in Cray Computer opened trading at around $14 in 1989, but dropped to $5 in early 1990.

The industry, meanwhile, was also in ferment over innovation. Parallel-processing computers could link dozens, hundreds, or even thousands of microprocessors to do big jobs. Not all problems would benefit from this, but for those that did, the machines could do the work "in less than one-tenth the time of a conventional supercomputer—and at about one-tenth the cost." It would be devastating competition. In response, Seymour is reported to have said, "If you were plowing a field, which would you rather use: two strong oxen or 1,024 chickens?" For those in need of less power, new competitors like Convex were now building small scientific computers, placing the market for large machines in further peril.

A prototype of the CRAY-3 was finally built. It was installed at the National Center for Atmospheric Research in Boulder,

Colorado, on a preliminary basis to run trials meant to identify and correct design bugs. But the company received no other orders for the machine. The market seemed to prefer "chickens," which did the same jobs more cheaply.

Cash ran low. The man who once received all the money he needed, almost without asking, was unable to secure additional financing. On March 24, 1995, Cray Computer filed bankruptcy. That same year, Cray Research, the company from which Cray Computer evolved, reported a 27 percent drop in revenue, a loss from operations of $50 million, and a restructuring charge of $187 million.

Conceding the market's judgment, Seymour set up a new company, SRC Computers (his initials), and started the design of his own massively parallel machine. He would attempt to reclaim his rightful position and prove that he still had ideas of both technical merit and economic appeal. At the age of seventy, he would begin anew. A wealthy friend agreed to invest. His new machine, Seymour said, would be "smaller than a human brain."

On September 22, 1996, Seymour Cray pulled his Jeep Cherokee onto Interstate 25 in Colorado Springs, Colorado. A thirty-three-year-old driver tried to pass but struck another car, then struck Seymour's jeep, causing it to roll over three times. The driver, unhurt, received a citation for careless driving. It took ninety minutes to extract Seymour from his jeep. He was taken to the hospital for emergency surgery, where doctors discovered he had massive head injuries and a broken neck. He never regained consciousness. Thirteen days later, on October 5, 1996, at the age of seventy-one, the great mind went forever quiet.

11

Conclusion

With the death of Seymour Cray the supercomputer era came to an end. His first company, Cray Research, had been acquired in February 1996, by Silicon Graphics for $30 a share cash for approximately 75 percent of the shares outstanding, and the remainder on a stock-for-stock exchange for Silicon Graphics stock selling in the $27 range. In 2000, Silicon Graphics then sold its Cray Research subsidiary to Terra Computer, which renamed itself Cray, Inc. Today Cray, Inc. operates as a public company based in Seattle developing high-performance computers. Revenues are running at about a third the level of its best years and the stock seldom draws attention.

Control Data, meanwhile, struggling to maintain profitability in the computer hardware market, diversified into many new areas. Already it had a huge financial subsidiary doing business in a wide range of markets. It had a large services business selling time on computers. It had a computer maintenance business. It had consulting services. It had an education arm. The peripheral products business had achieved a billion a year in revenues, much of it independent of the company's computer system sales. Still the company looked for new avenues of growth. It described its corporate mission as addressing "society's major unmet needs" and began to supply technical and financial help to many new businesses. One initiative, City Ventures, became a consortium to redevelop inner cities; another, called Rural Ventures, was set up to deliver technology to the family farm. There was also Renewable Energy Ventures and the Minnesota Seed Capital Fund. Control Data even started a new venture raising tomatoes hydroponically on the roof of one of its corporate buildings. Over all, investments were made in more than thirty small businesses. Almost nothing was outside its vision, as long as it offered an opportunity to address society's unmet needs while making a profit. In an effort to reduce recidivism among inmates at the prison in Stillwater, the company started a program to lease automobiles to newly released convicts so they could drive to work. But somehow cars got damaged, stolen, or otherwise disappeared. One car was used in a liquor store holdup that involved gunfire. These were not normal business problems.

In yet another response to social need, the company twice bought blocks of stock in Twin Cities companies to help prevent unwanted takeovers. This was temporary "white knight" behavior with no other expectation of benefit. In the process of doing these things, Control Data transformed its image from that of a computer hardware company into a company with noble aspirations but a puzzling business strategy.

Nevertheless, because of its many activities, revenues continued to advance, going from $3.7 billion in 1980 to $5 billion in 1984. Total employment reached 60,000 in 1981. In the same year, the company split its stock 2 for 1. On the newly split shares, earnings ranged between $4 and $4.50 a share for four straight years ending in 1983, but wary investors were willing to pay only an average of 8 to 12 times earnings, a huge contraction from the multiple of 100 times earnings when the company was young and looked like it could grow forever. Based on its recovered profitability and assuming its continuation, the company initiated a dividend in 1977 and increased it annually for seven straight years, hoping to establish a reputation of financial consistency and to provide a cash reward for its bruised shareholders. The computer business, having been bulked up by services offerings, was now three times larger than the financial business when measured by revenue, and roughly two times as profitable—finally achieving some of the proportionality expected by investors when Control Data and Commercial Credit combined some fifteen years earlier—but the financial reputation of the company was completely changed. In fact, from an investor's standpoint it looked more like the original Commercial Credit than Control Data, in that earnings were flat and the company paid a dependable dividend. There was also the attendant uncertainty of those unusual and unproven new ventures, but they were not a big part of the company's total revenue. The stock hit a high of $63 a share in 1983, which brought it to 15 times earnings, a reasonable valuation. That modest price, however, proved to be its highest for the rest of the decade.

In 1984, the company took a $130 million "special" restructuring charge against earnings, choosing to phase out the plug-compatible peripheral business and stating that the move was necessary because of pricing pressure on those lower performance products. The charge meant that earnings dropped 80percent. In its report to shareholders, the company said it was considering selling Commercial Credit "in order to concentrate more financial and management resources on its computer products and information services business," adding, however, that the sale was "not critical" since "the outlook is good." Norris was joined by two long-time employees in his comments in the annual report: Robert Price, who had been elevated to president, and Norbert Berg, promoted to deputy chairman.

In 1985, the company reported a devastating loss of $567 million. Computer business operations accounted for $143 million, partly because of a drop in revenue. Commercial Credit, surprisingly, reported a loss of $5 million because of increased underwriting reserves and a decision to discontinue the property and casualty insurance business. The remainder of the corporate loss came from "restructuring charges" for peripheral products and other operations. Liquidity became "a major problem." The company was in technical default on its bank debt, and Commercial Credit lost its investment grade credit rating. Given these reversals, the dividend was discontinued. By year end, Norris had turned seventy-five and announced his retirement as chairman, but remained a member of the board. For thirty-one years, he had overseen the evolution of a company he started without "really knowing what we were going to do." Over the years, it had become a major corporation employing many thousands; now it was struggling financially. Because of its lack of definition and its convulsive earnings history, most investors treated the company with indifference. Some, however, treated it with derision. They called the new business experiments "foo-foo" projects. *Inc.* magazine described Norris as an "eccentric corporate do-gooder." By year end, six members of the board of directors had retired, three of them because the company's director and officer liability insurance had been cancelled.

Strangely, Norris had become entangled in the same web of diversions he attributed to other managements back when his company was young and its mission clear. Asked at the time why he was successful and others in the industry weren't, he replied that they were involved in too many things and unable to concentrate on computer business matters that demanded prompt attention. Now he too lacked focus. His company seemed disoriented. Commenting on the pressures a company faces, Seymour Cray had concluded that most such diversions were brought on by investors wanting uninterrupted growth, not by management personalities or even economic reality. "I can't just say… I'm comfortable with this size company," he had said. On the other hand, surely he and Norris both knew that the absence of growth brings its own issue, usually slow deterioration. Seymour's company resisted the tendency to expand into new areas. Control Data did not. In the end, both companies were unable to overcome the countervailing trends in their basic businesses and suffered a similar fate.

Robert Price was named Norris's successor, taking primary responsibility for a company now struggling to rise out of a chasm of trouble. Assessing the situation, he stated that "operations had simply become too diverse," and that his objective was to "do fewer things more profitably... by focusing the organization on its core businesses." Because of adjustments that had already been made, total employment was down by 10,000. Price felt compelled to say, "We operate business ventures, not social programs." The stock fell to $15.

The financial cleansing of special write-offs sometimes means that the worst is over. But for Control Data, it wasn't. In the following year, the company lost another $264 million. Continuing to restructure, the company sold off "a dozen businesses or product lines." Write-offs associated with these sales, along with a $200 million "one-time restructuring charge," created most of the loss. The word *restructuring* had come to explain much of the computer division plunge into unprofitability over the previous three years. The annual report also stated that competition was "ferocious" in the peripheral products business, largely from Japanese companies. Three senior financial officers retired during the year, as did an executive vice president of technology and planning. Five members of the board of directors also retired.

In November 1986, Control Data completed a public sale of Commercial Credit stock and raised $530 million, while still retaining 18.4 percent ownership. The cash received from this sale, plus cash from the sale of other product lines and businesses, reduced the company's year-end debt to 50 percent of equity, down from 200 percent of equity just a year earlier in computer operations excluding the investment in Commercial Credit. This turnabout in the financial base of the company created a substantial improvement in underlying financial strength. But the consistent base of revenue from the operations of Commercial Credit that had sustained the company for almost two decades had been removed from the income statement. Control Data was again vulnerable to the irregularities of computer operations and the uncertainties of its other ventures. The company had undertaken this sale in an effort to bring liquidity to its badly shaken financial position.

A year later, the company sold its remaining Commercial Credit shares. This one-time gain, also called "restructuring," was sufficient to offset the expense of yet another restructuring reserve set up to absorb further losses in the disk drive business. It permitted the company to show a small profit for the year. Two more directors retired from the company's board.

In 1988, Control Data reported that two-thirds of its revenue generated profitable business, meaning that a third remained unprofitable, primarily the "computer mainframe and supercomputer businesses" and the "semiconductor activity"—this even though Cray Research was still performing at a high level selling supercomputers. Lawrence Perlman, an attorney by training, got recognized for his work on the sale of Commercial Credit and for restructuring the disk drive business, now a subsidiary named Imprimis, and was elevated to president. The company eked out a small profit for the year. Two more directors retired from the board, including Norbert Berg, an employee.

But consistent profitability remained elusive. In the first quarter of 1989, most U.S. employees went five days without pay, and executives took a 5 percent salary reduction for the first six months of the year. It wasn't enough. When the year ended, the company reported a loss from operations, and in addition took an enormous restructuring charge of $663 million, more than half of which came from closing down its supercomputer operation, now called ETA Systems. That division had sustained losses since being set up in 1983, and the company was unable to forecast with any confidence a better future. Imprimis, the disk drive subsidiary created in the process of many restructurings, was sold to a competitor, Seagate Technology, not for cash, but for a 17.6 percent interest in Seagate. These various restructurings reduced employment at Control Data by 15,500, and the attendant severance obligations contributed significantly to the company's loss. Three more board members retired, including John Buckner, who had been brought in as chief financial officer just three years earlier to oversee much of the restructuring effort. The company again fell out of compliance on its credit agreements, but the banks granted waivers. Price and Perlman, chairman and president, respectively, stated in their year-end report that "the first priority for Control Data Corporation is to achieve consistent and growing profitability." Most shareholders probably nodded in agreement. It had been a period of repeated disappointment and interminable struggle.

The year 1990 brought improvement, but not much. Again the company took some charges against earnings, this time to cover cost overruns as well as restructuring activities. Earnings at $2.7 million barely exceeded break-even. The company sold its Seagate stock plus some other holdings and generated $153 million in cash. Lawrence Perlman was now chief executive officer, as Robert Price had retired. The company's revenues were $1.7 billion in 1990, down from a peak of $5 billion seven years earlier. Much of the decline was due to the elimination of unprofitable businesses, but ongoing profitability was still not consistent or dependable. The corporate aggregate that Norris had built and nurtured over many years had undergone multiple amputations. "When they [broke] up the Univac Division, part by part, limb by limb," he had said earlier in his career, "I left." Now he was seeing his own creation suffer a similar fate, although from larger economic forces. He decided to take final leave of the company and chose not to stand for reelection as a board member at the next annual meeting. It was a quiet departure. Another director retired also.

Years later, when celebrating the company's fiftieth anniversary, Ed Orenstein, who sold his company, Data Display, to Control Data in 1965, said, "Norris's social conscience was remarkable, but not completely compatible with running a growth company. [The company] would have been better off if he had stepped aside [earlier]." He spoke for many. Commenting at the same time, Lawrence Perlman said, "I think the speed of the technology change really took a lot of people by surprise," and added that pet projects got attention, while many mainstream businesses didn't get enough attention.

Declining sales in computer systems led to a loss of $9.8 million in 1991. The company took more restructuring charges, and in the process chose to become an "open systems integrator" for computer systems—that is, one incorporating competitors' products, when appropriate, to create the best system for a given customer. It was a practical decision, recognizing the sometimes superior technologies of other manufacturers. By year end, the company was working on plans to "reshape" itself in order to "unlock" value. Perlman stated that "the more sharply focused a business, the better its chances for success." Annual revenues were now $1.5 billion. The stock traded as low as $7 a share in 1990 and 1991

In 1992, the much diminished company was divided into two entities: Control Data Systems (a computer systems integrator) and Ceridian (information services and defense electronics). Reflecting the increased emphasis on services over the years, Ceridian had revenue three times larger than Control Data Systems and was considered the successor company. Shareholders received one share of Control Data Systems for every four they held in Ceridian. Five years later, Control Data Systems, Inc.

was acquired by the New York investment firm Welsh, Carson, Anderson & Stowe for $20.25 per share. It was sold again in 1999 to a British firm, BT Global Services, and with that final transaction the simple, two-word company name originally created by matching possibilities on a piece of paper—a name that quickly acquired a golden aura and gave ordinary people enough wealth to buy houses or retire in comfort, a name that represented extraordinary technical achievement and gained customers all over the world, a name that electrified Wall Street—was retired from public use. The business was renamed Syntegra.

But shareholders still owned Ceridian. In 1999, the stock split 2 for 1. The company was making money and growing, but had no clear industry definition. It sold one division, split off another, and made various supplemental acquisitions. In 2007, it too was acquired, at $36 a share—a respectable increase from the worst days of its predecessor. But, figuring in the two 2-for-1 splits since the stock hit its peak in 1968, each current share brought $144 (calculated at 4 times $36), plus the Control Data Systems payment of about $10 a share (2 times $5.0625, before the second Ceridian split), and additional amounts for various shares previously spun out to shareholders, plus dividends. The total was close to $174, its earlier peak. But it took nearly four decades to recover that amount. Few remembered that the stock came public at $1 a share and that it went to $100 in four quick years, because that initial offering took place mostly among friends and acquaintances a distant fifty years in the past. Instead, they seemed to remember the recent diversions into unusual ventures, the repeated write-offs, and the company struggling, and struggling some more, to make a comeback.

There was no funeral for the passing of Control Data from public ownership, no eulogy of praise beyond the normal newspaper reports. No one played taps for what had once been a national resource. The company had given way to new forces of life, similar to the relentless cycles of nature in which organic matter inevitably dies, decomposes, and helps support new life. "Creative destruction," Joseph Schumpeter, the Harvard economist, called it. The forces, he said, were a "perennial gale," and brought benefit to the general population through new inventions, services, and methods of delivery. In the early days, those supportive winds allowed Control Data to soar, to be part of the gale, to help form the new creation. But as it advanced in age, the company became caught in a dead zone and began to lose altitude. The gale had shifted. Other companies were in ascendance, maybe even creating their own uplift.

"Trees don't grow to the sky," they say on Wall Street, and they begin to search for new shoots.

Not everyone held on until the final liquidation. In 1967, Ray Kraushur looked over his investments and decided to sell Control Data stock at a price close to the top. He then reinvested some of the proceeds in Medtronic, managing to leave one trapeze in mid-air and grab another in a bravado maneuver based on either brilliance or luck. No formulas, charts, or professional advisors guided his decision. Instead, he did personal research within his own community of friends and used common sense. At the time he was a barber cutting hair in downtown St. Paul for $3 a customer. The gain he took on Control Data stock, he said, "represented a lot of haircuts."

He had been a marine for five years, serving in the Pacific during World War II. Following that he began barbering, saved his money, and was able to get his own shop in 1952, after which he remembers having $3.75 left to his name. In the late 1950s, another former marine became a regular customer and repeatedly urged him to buy Control Data stock. At first, Kraushur wasn't comfortable with the idea, but he finally did as suggested, largely to please his new friend, who already owned the stock.

Twenty years later his success was featured in the book, *Moneymaking.* No numbers were given, but it said that he then owned a commercial building, a portfolio of investments that included bonds, his own airplane, and three cars, apparently all from cutting hair, listening carefully, and investing wisely. On weekends he liked to fly to northern Minnesota for fishing trips. Still, he said, he continued to enjoy cutting hair. He may have had life's comforts figured just about right. And he seemed to have a sensible investment touch.

Control Data wasn't the only company to spring from Univac, just the most prominent. David Lundstrom remembered a meeting where Univac alumni—called Unihogs because they met every year on Groundhog's Day—created a genealogy listing seventy-six firms emanating initially from Univac and evolving in some cases five generations; and the members weren't sure they were done counting. Univac was almost a solar explosion of new business in the Twin Cities community.

Although Control Data, the primary meteor to spring from Univac, eventually lost its luminance, it too gave off sparks that became separate companies. Cray Research was certainly the most prominent. Former president Bob Price, in his book on innovation, cited sixteen companies started by former Control Data employees. This didn't include the assets spun out by Control Data or the companies started with Control Data funding and assistance, which would easily add dozens more to the list. In short, Univac and Control Data had an impact on local industry far greater than their individual successes. But the passage of time and the evolution of technology eventually moved the focus of the computer industry to the West Coast, where it resides today.

Still, Minnesota was not left with empty offices and idle workers. A whole new medical industry had taken shape. "I think the reason [medical] devices were so strong here was that we had the computer industry as a forerunner, and the base of many devices is technology," said Norm Dann. He had an engineering background, became a Medtronic executive when his business was acquired, and then became a principal at Pathfinder Venture Capital, a Twin Cities fund that invested in computer and medical technologies. He was in a position to know.

The new industry was perhaps best represented by two big publicly traded stocks: Medtronic, the world's largest medical device company at $16.5 billion in revenue in 2011, and St. Jude Medical at $5.6 billion revenue. 3M (Minnesota Mining and Manufacturing) had a health care division doing approximately $5 billion in annual revenue. Patterson Companies did $3.5 billion in revenue, mostly in dental supplies. The Cardiac Pacemaker business continued to be significant in the area also, first as part of Guidant and more recently as part of Boston Scientific.

Like the computer industry earlier, many smaller medical companies came into existence because of the thriving environment—up to 350 companies, according to some sources. A review of the *Star Tribune 100* brings up the following publicly traded medical companies in 2012.

Techne	$306 million, revenue in 2011
Medtox Scientific	$108
Vascular Solutions	$90
Cardiovascular Systems Inc	$80
Surmodics	$67
Rochester Medical	$56
MGC Diagnostics	$29
Electromed	$20
Uroplasty	$19
Urologics	$14

Also in the area is United Health Group, a giant health benefits supplier listed on the New York stock exchange.

In the suburb of Arden Hills, the former Control Data manufacturing plant at 4201 Lexington Avenue is now home to the cardiac rhythm management (CRM) business of Boston Scientific. The fourteen-story headquarters building that Control Data built at the intersection of Old Shakopee Road and 33rd Avenue in Bloomington (later home to Ceridian) is now occupied by HealthPartners, a large health care provider. It is as though the new industry ordered the old one to move over. Ceridian, now privately owned, is across the road.

In 2011, a couple of men were talking at the Medtronic annual meeting. One said, "You know, I really don't have to work for a living. But I would rather people didn't know it."

"Really, how's that?"

"My father bought Medtronic stock back at the beginning. I have it now. It's enough to live on."

"Isn't that nice? Was he an employee?"

"No. He worked at Honeywell on thermostats. Cut his wrist one day on some extremely thin sheet metal. When he got the wound treated, the nurse mentioned Medtronic stock, so he bought 100 shares at $5. When it dropped to $3.50, he thought he might have made a mistake."

"Uh huh."

"He bought more at higher prices and sold some too. When he died, we had to sell some of the stock to pay $240,000 in estate taxes. Basically I inherited the original 100 shares. I am an only child, see."

"Did your father ever enjoy any of the benefits?"

"He retired at age sixty. Lived well. The dividends were enough to buy a new car every two or three years."

"Let's see," said the listener, "100 shares turned into . . ." and he checked the history of splits. It took him a minute or two to do the arithmetic.

Split 2 for 1, August 1967, 100 shares became 200 shares
Split 2 for 1, January 1969, 200 shares became 400 shares
Split 2 for 1, September 1972, 400 shares became 800 shares
Split 2 for 1, July 1980, 800 shares became 1,600 shares
Split 2 for 1, August 1989, 1,600 shares became 3,200 shares
Split 2 for 1, August 1991, 3,200 shares became 6,400 shares
Split 2 for 1, September 1994, 6,400 shares became 12,800 shares
Split 2 for 1, September 1995, 12,800 shares became 25,600 shares
Split 2 for 1, September 1997, 25,600 shares became 51,200 shares
Split 2 for 1, September 1999, 51,200 shares became 102,400 shares

"One hundred shares became 100,000 shares. Wow!"

"Yup."

"Dividends are 97 cents a share. That means $97,000 a year for you. And they've been going up every year."

"That's about right. But I'd rather people didn't know about it."

Notes

Chapter 1: The Birth of a Champion

Information on the formation of ERA, the history of Control Data, and the life of William Norris are taken principally from the ten sources listed below:

James C. Worthy. *William C. Norris: Portrait of a Maverick.* Ballinger, 1987.

Carol Pine and Susan Mundale. *Self-Made: The Stories of 12 Minnesota Entrepreneurs.* MCP, 1982. Chapter six features Norris.

Don W. Larson. *Land of the Giants: A History of Minnesota Business.* Dorn, 1979. See chapter 11, "The Computer Age."

Ralph Nader and William Taylor. *The Big Boys: Power and Position in American Business.* Pantheon, 1986. The chapter on William Norris includes quoted conversations that took place when Cray joined Control Data.

Erwin Tomash and Arnold A. Cohen. "The Birth of an ERA: Engineering Research Associates Inc. 1946 – 1955." *Annals of the History of Computing*, Vol. 1, No. 2, Oct. 1979.

Bill Hakala and Associates. *Engineering Research Associates: The Wellspring of Minnesota's Computer Industry.* Sperry, 1986.

David L. Boslaugh. *When Computers Went to Sea.* Wiley, 1999.

David E. Lundstrom. *A Few Good Men from Univac.* MIT, 1987.

Charles J. Murray. *The Supermen: The Story of Seymour Cray and the Technical Wizards behind the Supercomputer.* Wiley, 1997.

Arthur L. Norberg. *Computers and Commerce.* MIT, 2005.

I also drew upon oral interviews at University of Minnesota Babbage Institute, especially those of Willis Drake, Arnold Ryden, William Norris, Frank Mullaney, and John Parker. An interview with Seymour Cray at National Museum of American History, Smithsonian Institution, was also useful. Unless otherwise noted, stock prices throughout the book are drawn from local over-the-counter stock listings in local newspapers bound into volumes at the downtown Minneapolis Hennepin County library.

Norris's early life was described in considerable detail in an article in the *World-Herald Magazine,* dated December 22, 1963.

Early Control Data history was summarized well in "Control Data, from Mystery to Legend," *The Upper Midwest Investor* (November 1961).

The remark about a "seething resentful atmosphere" will be found in the oral history of Willis Drake. ERA net worth of "$150,000" comes from the Ryden interview, debt "twice that" from *The Supermen,* p. 33. The "bucket of cold water" quote will be found in Willis Drake's oral history interview Information about the investor presentation at the Twin Cities Town and Country Club comes from a conversation with Dick Slade, who arranged the meeting.

Quotes from the newpapers cited were obtained from files of clippings concerning Control Data and William Norris found at the University of Minnesota Babbage Institute. Wheelock Whitney's comment is from the book *Moneymaking in the Twin Cities Local Over-the-Counter Marketplace* by James R. Ullyot, 1972.

The book *Buffett; The Making of an American Capitalist* by Roger Lowenstein, 1995, mentions Buffett choosing not to invest in Control Data, page 68. Comments concerning other potential investors were from conversations with knowledgeable participants.

The comments on Willa Cather were confirmed in conversation with Connie Van Hoven, William Norris's daughter. She also provided information on the Buffett relatives and on her parents' marriage.

Comments on Norris are based on observations made by the author while an employee of Control Data, also common folklore, submitted for review to two of his children.

The reference to Willa Norris as Vice President of Mingling appears in an obituary that appeared in the *Star Tribune* on July 24, 2006.

The quote from Sid Hartman's sports column can be found in the book: *Sid!* by Sid Hartman with Patrick Reusse, 1997. It originally appeared in the July 30, 1957, edition of the *Minneapolis Tribune.*

General Mills information is taken from *The Upper Midwest Investor,* November 1961 issue. Various issues of the magazine were made available from the personal collection of James C. Fuller, editor. These and other miscellaneous sources provided information on other technology companies in existence at the time.

Atlas computer comments confirmed in conversation with Doug Larson, a former employee of ERA and Control Data.

List of early employees at Control Data came from a document in the possession of Dean Laurance, a former Univac and Control Data employee who knew the people. Arnold Ryden is technically listed as employee number one, and Bill Drake number two, but various articles discussing those times refer to Bill Drake as employee number one. William Norris is technically listed as employee number five even though he was the unquestioned leader. Laurance says that Bob Kisch told him about discovering that Frank Mullaney had skipped two grades in school.

Comments on Cray's personal mannerisms are based on the author's observations at annual meetings and other occasions, and confirmed by fellow employees of Seymour Cray, especially Les Davis, Pete Zimmer, Doug Larson, Mike Schumacher, and Sam Slais, who worked with him, and in later years, Greg Barnum. Other details about his life provided by the sources listed above and in "Reclusive Genius, Seymour Cray," excerpts from a speech given by him on Nov. 15,1988, reported in *Computerworld,* June 22, 1992.

"The Genius," the cover article from *BusinessWeek*, April 30, 1990, contains much information on Seymour Cray's personal life, as does "Meet Seymour Cray," an article in *Twin Cities Reader*, August 31, 1983.

Professor Cartwright's comment on Seymour Cray was recalled by his son, Paul O. Cartwright, who worked at Control Data.

An interesting subnote on Seymour Cray's abilities: Cray wrote a basic operating system for the Control Data 6600 machine. A team of programmers in California was responsible for enhancing it with a full complement of features, but after months of effort had to admit failure. The company went back to Seymour's basic package and standardized it as an official product offering. With that, Seymour became recognized as not only a top mathematician and engineer, but also very likely the company's best programmer.

Control Data Corporation results, accounting information, and comments are taken from the initial prospectus and annual reports available at the Charles Babbage Institute. Because of acquisitions, the numbers in annual reports were restated for previous years, but amounts cited here are from actual year-end reports (unadjusted) that were compiled in a *Financial History* by Marvin Rogers, Bjarne Eng and Jack Karnowski, financial officers of Control Data Corporation, available in the Babbage Institute. The "double declining balance" depreciation method was described on page 15 of the *Financial History.*

"Big Datty" reference was provided in an interview with Howard O'Connell, local securities firm executive at the time. Comments about Governor Perpich come from papers provided by Norbert Berg. Both Berg and Perpich were employees of Control Data. The George Waters comment is from the collection of oral historoes at the

Minnesota Historical Society, *Pioneers of the Medical Device Industry in Minnesota, An Oral History Project.*

Chapter 2: Local Market, 1957–1959

The author visited Minneapolis and St. Paul about once or twice a year in the 1950s, attended the University of Minnesota as a graduate student in finance in 1959 and 1960, and worked at the First National Bank of Minneapolis as a security analyst from 1961 to 1963. Many of the observations are from his personal recollection and from his knowledge of financial markets. Drafts of this material were reviewed by people who lived in the Twin Cities at that time and by those who worked in investments. Their comments were incorporated. At the time, the author also observed the New York Stock Exchange at work.

Concerning stock trading in increments of an eighth, "Spanish dollars in use in the colonies had to be cut in halves, quarters, and eighths, called bits, to make small change." From an article by economic historian John Steele Gordon in *Barron's* newspaper, July 9, 2012.

Names of brokerage firms in business at the time came from personal memory of the author and the recollection of others who worked in the industry and from advertisements in publications from that time. The average age of brokers came from the memory of Howard O'Connell, who had access to Minnesota Securities Division records as an employee. The quote is recorded in *The First Fifty Years,* a pamphlet about John G. Kinnard and Co.

Tax rates are mentioned in *The Upper Midwest Investor,* March 1962. An investor with a taxable income of $16,000 was in the 47 percent federal income tax bracket.

The "cops warning" comment happened to the author's uncle.

Gandy dancers information provided by Jerry J. Johnson, who worked occasionally in his father's bar on Washington Avenue, and a brochure from the Gandy Dancer Trail in northwestern Wisconsin, which states that the name described hand crews who built and maintained railroad tracks with tools manufactured by the Chicago-based Gandy Tool Co. "The crews used vocal and mechanical cadences to synchronize the swinging of their hand tools or the movement of their feet. Hence the name Gandy Dancer."

Comments on Charlie's Café Exceptionale came from *Minnesota Eats Out* by Kathryn and Linda Koutsky, supplied by *Star Tribune* reporter Ann Burckhardt.

Comments on the Waikiki Room based on author's own experience. Information on Minneapolis Honeywell came from an article in

Electronic Week, February 17, 1958, and from *The Upper Midwest Investor*, November, 1961. Stock price history is from a chart from that period, which also cites the date of the name change.

Background on Minnesota Mining and Manufacturing derives from the book *Brand of the Tartan* by Virginia Huck and an article in *Barron's* newspaper (February 28, 2011). Stock history on Minnesota Mining and Manufacturing was provided by a 3M archivist.

Minneapolis Honeywell Regulator stock information came from a chart of that period. I derived Investors Diversified Services stock history from newspaper quotes compiled by the downtown Minneapolis Hennepin County Library.

Chapter 3: Hot Stocks

New listings were derived from a comparison of local over-the-counter stock listings from local newspapers found in bound volumes at the downtown Minneapolis Hennepin County Library.

Information on founders of Data Display comes from *The Upper Midwest Investor* (December 1961) and from the book *A Few Good Men from Univac* by David E. Lundstrom, 1987.

I obtained information about Les Bolstad Jr. and Pat McNeil in personal interviews. I also interviewed Eddie Fleitman and got further informaton about him from George Bonniwell.

Product Design and Engineering facts came from *The Upper Midwest Investor* (April 1961). The facts on the Convention Center's new issue also came from *The Upper Midwest Investor.* Univac employee points of view are taken from *A Few Good Men from Univac* by David E. Lundstrom, 1987. Rocket Research information is derived from *The Upper Midwest Investor* (May 1961).

Background information about Twin City brokerage firms comes from reminiscences of those involved at the time, plus ads in the *Upper Midwest Investor.*

Information about the financial requirements of brokerage firms comes from conversations with Howard O'Connell and John Steichen. The comments of George Kline and Jim Fuller are based on interviews. Fuller provided me with copies of the magazine he published, *The Upper Midwest Investor,* as general source material.

The references to John Stephens and Co. are based on an interview with Stephens. Robert Smith's comments will be found in the article "Market Outlook" in *ACE* magazine (November 1961). Nucleonic Controls information comes from a prospectus provided by Howard O'Connell.

Remarks by Commissioner Arthur Hansen appear in *The Upper Midwest Investor* (June 1961). Willis Drake comments come from the *The Upper Midwest Investor* (April 1961).

Arnold Ryden information is found in files at the University of Minnesota Babbage Institute and the downtown Minneapolis Hennepin County library business files. Especially helpful was an article in the *Minneapolis Star* (June 17, 1959).

The Telex company described in *Moneymaking* by James R. Ullyot, p. 16, also *The Upper Midwest Investor*, December 1961.

Norris's comment about Ryden is from a *Forbes* magazine article, August 1, 1962. Jim Secord's comments are based on an interview.

The Continental Securities report was provided by Howard O'Connell.

Chapter 4: Popular Delusions

The Norman Terwilliger remark can be found in *The Upper Midwest Investor* (July 1961). Details of 3M history come from *Brand of the Tartan.*

Bernard Baruch's comments come from his autobiography *My Own Story*, 1957. Examples of historic excesses can be found in *Extraordinary Popular Delusions and the Madness of Crowds* by Charles Mackay. Stock recommendations are from a Continental Securities broker report to customers, "What's Ahead for the Market in 1963?" in Howard O'Connell's personal file.

Twin Cities electronic industry statistics are from a guest column by Merrill J. Anderson in *The Upper Midwest Investor (*May 1961). Statistics comparing market size are taken from *Mid-American Investor* magazine—the follow-on publication to *The Upper Midwest Investor* (February 1963).

Information on Possis Machine is taken from an article in *The Upper Midwest Investor* (May 1961) and a *St. Paul Pioneer Press* profile of the company dated January 16, 1984. Its stock prices can be found in the annual publication of *Corporate Report*. A commemoration in the *Possis Corp.* annual report, dated 1993, summarizes the career of its founder. I also obtained information on the company from company annual reports, *Corporate Fact* books, and *Moneymaking in the Twin Cities Local Over-the-Counter Marketplace*, by James R. Ullyot, 1972, pp. 31–35.

Initial offering price for Tonka Toys appeared in the March 1962 issue of *The Upper Midwest Investor.* Company comments are from *The Upper Midwest Investor* (September 1961 and June 1962).

Chapter 5: Cinderella Grows Up

Along with sources cited earlier, I drew in this chapter on a number of articles in national publications describing the firm's sudden rise to prominence, including "Stock Continues to Soar" (*New York Times,* August 30, 1963); *Forbes* (June 1, 1964); "Poor Man's IBM," (*Time,* August 14, 1964); "What Makes a Growth Company?" (*Dun's Review,* October 1964); "The Sweetest Stocks of '67" (*Newsweek,* December 25, 1967).

A comparison of Control Data, Xerox, and Polaroid will be found in *Minneapolis Star* (August 26, 1963).

The famous memo describing IBM's Thomas Watson Jr. as "furious" was quoted in the *Wall Street Journal* (May 19, 1982) and repeated in several books on the era.

For a list of Control Data's acquisitions by year, see *Financial History* by company officers, final appendix. A description of its change in accounting policies can also be found in *Financial History.* This change is also discussed in "Is Control Data in Trouble?" (*Forbes,* September 15, 1965). See also "Control Data a Disappointment?" (*Minneapolis Tribune,* June 20, 1965).

Dick Jennison's praise of Control Data appears in a memo dated August 4, 1965, available at the University of Minnesota Babbage Institute.

Norris's comment about "butts from third base…" appears in "Is Control Data in Trouble?" (*Forbes,* September 15, 1965). Norb Berg said in an interview with the author that Norris told him he really said that they didn't know their "ass from third base about the computer business."

The comment on first-quarter earnings comes from the *Wall Street Journal,* September 22, 1965. Mullaney's resignation was reported by the *Minneapolis Tribune* (February 9, 1966) and *Electronic News* (February 14, 1966), in an article that includes a denial by Seymour Cray that he might resign.

For reports on Kisch and other resignations, see summary article from *Minneapolis Star* (July 21, 1966). More summary information appears in *Wall Street Journal* articles dated July 21 and 22, 1966. Perhaps the most comprehensive summary can be found in an extensive *Wall Street Journal* article dated August 4, 1966. A Norris explanation of the resignations can be found in the Control Data 1966 annual report.

Computing Newsline, dated February 23, 1966, discussed the reasons for resignation by Mullaney and Kisch. Control Data "survived the worst" is from *Electronic News* (February 21, 1966). See also the *Minneapolis Tribune* and *Star* editions dated February 16, 1966.

"Genius taken for granted" quote about Cray is from *Fortune:* "Control Data's Magnificent Fumble" (April 1966). "He almost ripped the place apart" quote appears in *Fortune:* "Control Data's Newest Cliffhanger" (February 1968). The phrase "Like a dam breaking" appears in "Control Data Rebound Shown in Stock Leap" (*Minneapolis Tribune,* July 30, 1967).

Harold Hammer's background information is from *Financial History,* page 155. The acquisition of Commercial Credit is described in the appendix of *Financial History* and also in the *Minneapolis Tribune* (June 17 and August 18, 1968) and the *Wall Street Journal* (July 2, 1968). The early history of Commercial Credit can be found in the Control Data 1972 annual report.

Chapter 6: A Sleeper

I obtained information on Earl Bakken, Palmer Hermundslie, and the early days of Medtronic from the following sources:

One Man's Full Life by Earl Bakken, 1999.

The Story of Medtronic, pamphlet published by the company.

The Mission, Celebrating 50 Years, Medtronic, by Nancy Blakestad, 2011.

Self-Made: The Stories of 12 Minnesota Entrepreneurs, by Carol Pine and Susan Mundale, 1982, chapter 2.

I also found useful an initial prospectus obtained from Thomas Holleran, first legal counsel to the company, board member, and later president; Medtronic annual reports; and visits to the Bakken museum in Minneapolis.

References to "blue babies" come from various sources, especially *King of Hearts: The True Story of the Maverick Who Pioneered Open Heart Surgery,* about Dr. C. Walton Lillehei, by G. Wayne Miller. The book says that the death of a Lillehei patient from a power failure in October 1957 was a "rumor," but both Bakken and Holloran said it actually happened. Additional information came from *Pioneers of the Medical Device Industry in Minnesota, An Oral History Project* from the Minnesota Historical Society, 2002.

Historical information on pacemakers comes from the following Internet sources that cross-verify the basic facts:

Who Invented the Pacemaker?

Pioneers of Cardiology, Journal of American Heart Association, June 5, 2007.

Health Technologies Timeline
The Magnificent Century of Cardiothoracic Surgery: Reviving the Dead, Heart Views, Jun–Aug 2010.
Artificial Pacemaker

The seventy-two-year-old man who received the first permanent Medtronic pacemaker lived another six years according to *The Mission* and seven years according to *One Man's Full Life.*

Debenture offering information is taken from Medtronic prospectus dated December 18, 1959. The comments from Tony Gould and Tom Holloran were obtained from personal interviews. Chardack Greatbatch pacemaker information comes from *The Mission.* The *Saturday Evening Post* article discussed in *The Mission* was dated March 14, 1961.

Company information in 1961 was taken from *The Upper Midwest Investor,* October 1961 and a sidebar in the September 1961 issue.

Information about company difficulties in 1962 comes from various sources. It was reviewed by both Earl Bakken and Tom Holloran for accuracy. Holloran remembers that Mallory offered to pay approximately $1 million for the company, but *The Mission* (p. 48) says $3 a share. With 413,000 shares outstanding on April 30, 1962, the $3/share price would be a little over $1.2 million, plus the assumption of $323,000 in debt.

I obtained important information about Earl Bakken's dancing ability from his classmates Pat and Mary Ann McNeil. Facts on his remarriage come from *Self-Made: The Stories of 12 Minnesota Entrepreneurs.*

Board of directors comments to Earl Bakken on executive leadership were taken from books on Medtronic and expanded in conversation by Tom Holloran, who was there. Comments on Manny Villafana were taken from various sources and were confirmed in an interview with him.

The number of pacemakers shipped in the United States today is taken from a *Wall Street Journal* article dated October 4, 2011. The worldwide number is taken from an editorial article in the *Star Tribune,* April 14, 2011.

Chapter 7: A Meteor

General information about Quarterback was derived from annual *Corporate Report* books at the Minneapolis Hennepin County Library and from the book *Moneymaking.* The conversations in this chapter are imaginary, but the prices cited are genuine.

Comments by Mike Gruidl and Jim Fuller come from personal

interviews. The listing of stocks and their price changes from December 27, 1966, to December 28, 1967, along with accompanying quotes came from the book *Moneymaking*, which was quoting from the magazine *Commercial West*.

Comments on "blind pools" come from an interview with George Kline. "Regulation A" requirements are mentioned in the book *Moneymaking*, pp. 73 and 83.

Retired *St. Paul Pioneer Press* columnist Dave Beal gave me copies of his articles from the *Milwaukee Journal*, October 5–10, 1980.

Information on Pentair's current status are taken from *The Value Line Investment Survey*.

Information on stock trading limitations in the late 1960's came from a *Barron's* newspaper article by historian John Steele Gordon, dated August 27, 2012, and also the John G. Kinnard & Co. publication, *The First Fifty Years*. Current trading numbers are available in both *Barrons's* and the *Wall Street Journal*. The surge of orders for Control Data computers is remembered by the author, who was an employee at the time. Verification came from the mention of certain orders in the Control Data annual reports of 1968 and 1969.

The Minnesota Securities Department workload was mentioned in a Dave Beal article.

Chapter 8: Success and Hard Times

For references to "paper machines," see the *Wall Street Journal* (December 12, 1968).

Information concerning the IBM lawsuit comes largely from a report titled "History of CDC v. IBM" written by Elmer B. Trousdale of the law firm Oppenheimer, Wolff, Foster, Sheppard and Donnelly, (July 1980), available at the Babbage Institute, University of Minnesota. The author was unable to locate any comparable source from the contending law firm, and is aware his material represents only one side of the case. The basic facts are not in dispute, and in general are supported by newspaper accounts at the time and Control Data annual reports and by comments in the book *A Few Good Men from Univac* by David Lundstrom, p. 173.

The *Wall Street Journal* (December 12, 1968) quoted an IBM spokesman as saying the suit came "as a complete surprise." Various valuations of the Service Bureau Corporation at "$50 to $180 million" can be found in *Financial History*. The description of the Service Bureau Company business taken from the Control Data's 1972 annual report.

The practice of "bundled" pricing and its effect on the industry is described in "History of CDC v. IBM" referred to above. Norris's comments are taken from Control Data's 1969 annual report. Subsequent effects of bundling come largely from comments in Control Data annual reports. Control Data executive Bob Price said in a *Star Tribune* interview (October 12, 2007) that "Microsoft owes its existence to unbundling."

The quote of a former high-level executive (unnamed) about Norris is taken from *The Big Boys: Power and Position in American Business,* p. 459. The quote concerning "Franklin Roosevelt reforms" is taken from *Self-Made.* The quote from Norris about "compulsion to go south" is from the *Minneapolis Star Tribune* (January 9, 1986).

The description of company activities in the 1970s is taken from annual reports in the years cited, *Financial History,* and from an interview with Norbert Berg.

Seymour Cray's resignation and the word *appalling* are mentioned in the book *A Few Good Men from Univac,* which also notes that trading was suspended in the stock.

Cray's reference to receiving early support from Norris appears in the *St. Paul Pioneer Press* (December 7, 1981), which also reported "There really had to be a separation." Development of the STAR-100 computer is described in the 1973 Control Data annual report.

The 55-mile-per-hour speed limit was part of the 1974 Emergency Highway Energy Conservation Act.

Stock prices and related information on local firms in this chapter were taken from annual volumes of *Corporate Report,* especially the years 1974 and 1975.

A report from the Minneapolis investment community comparing current stock prices to the Great Depression probably was written by Steve Leuthold or Ed Nicoski, both market commentators at the time. In interviews, both said it sounded like something they might have done, but neither could verify it. RCA stock was mentioned in the book *The Great Crash, 1929,* and other historical references about that time. The "recurring" quotes came from John Kenneth Galbraith, author of that book.

Chapter 9: The Entrepreneur

Information on the early life of Villafana and his later medical ventures came from the following sources:

Self-Made: The Stories of 12 Minnesota Entrepreneurs, chapter 9.

Pioneers of the Medical Device Industry in Minnesota, An Oral History Project, pp. 529–559. The same source has interviews with Anthony J.

Adducci, Earl Bakken, C. Walton Lillehei, Demetre M. Nicoloff, and Thomas E. Holloran, all participants in early medical innovations in Minnesota, which proved helpful.

Both Villafana and Holloran were interviewed by the author to verify various facts taken from a variety of sources.

Facts concerning Cardiac Pacemakers, Inc., were found in the initial offering circular for common stock, dated May 26, 1972, provided by Mr. Villafana, and in copies of the company's annual reports. Stock prices come from annual *Corporate Report Fact Books*. The Greatbatch quotes come from *The Making of the Pacemaker,* Wilson Greatbatch, 2000, pp. 140–141.

Terms of the buyout from Eli Lilly & Co. are reported in *Moody's OTC Industrials* (August 18, 1978). Cardiac Pacemakers was spun out as Guidant medical device business in 1997, which was later bought by Boston Scientific/Abbott Labs.

Villafana's remark, "Many times I felt like giving up," comes from a report nominating Villafana for the Minnesota Business Hall of Fame (September 30, 1983).

A conversation between Dr. Demetre Nicoloff and Zinon (Chris) Possis appears in *Pioneers of the Medical Device Industry in Minnesota.*

Information on St. Jude Medical, Inc., comes from the initial stock prospectus dated February 9, 1997. Frank Hurley made his summary comment during a personal interview. I obtained information on GV Medical from company annual reports and from *Corporate Report Fact Books.*

Information on Helix BioCore is summarized nicely in an article in the *Star Tribune* (December 9, 2004). The article also refers to ATS Medical. The CABG Medical stock offering was covered in the *Star Tribune* (December 8 and 9, 2004). The "Device not working" comments appear in the *Star Tribune* (August 6, 2005).

The decision to close CABG Medical was covered in the *Star Tribune* (February 9, 2006). Villafana is "deeply saddened." The *Star Tribune* also covered the meeting on CABG Medical (April 28, 2006). It reported that liquidation of CABG approved by "98%, but a shareholder 'grumbled'." Terms of the liquidation also appear in *CCH INCORPORATED,* Capital Changes Reports (3210: 8-3-2012)

Ron Thomas's comment comes from an interview with the author. Kips Bay Medical quotes are from an article in the *Star Tribune* (November 4, 2008). Villafana says "we'll create a new industry." Dan Carr says Villafana is "the Cardiac Kahuna."

Kips Bay Medical's initial stock offering was described in the *Star*

Tribune (February 12, 2011). I found details of the underwriting in *Moody's*.

Chapter 10: Cray Research

In addition to material about Cray mentioned earlier, I found an article titled "The Genius" from *BusinessWeek* (April 30, 1990) especially helpful. Details on financing, revenue, and earnings came from company annual reports. Stories about Cray and John Rollwagen derive from interviews by the author with Rollwagen.

Comments about early employees at Cray Research come from a telephone interview with Les Davis, who was one of them. The "world's most expensive love seat" line appears in *Twin Cities Reader* (August 31, 1983). The six-month evaluation report from Los Alamos is discussed in the company's 1976 annual report. The "Deep thought" comment appears in the *Wall Street Journal* (April 15, 1979). Threatening phone calls are mentioned in the article "The Genius" cited above.

"Plans to decrease reliance on [Cray]" are reported in the *Minneapolis Star* (May 10, 1979).

That the stock price is "greatly inflated" is quoted in the *Minneapolis Star* (May 9, 1980). The article also mentions "two year slippage" in chips, and emphasis on lease revenue.

The Morgan Stanley quote on earnings estimates appears in the *Minneapolis Tribune* (September 6, 1980) repeat of an article from the Dow Jones News Service.

Listing on the New York Stock Exchange is mentioned in the 1980 annual report and also the *Minneapolis Tribune* and the *Pioneer Press*, (both dated November 20, 1980). *Fortune 500* listing is mentioned in the 1987 annual report, on the company's fifteenth birthday.

The halt in trading in November 1981 and the subsequent announcement is covered by both the *Minneapolis Tribune* (November 19, 1981) and the *St. Paul Pioneer Press* (November 19 and 20, 1981). The *Pioneer Press* provided more information three days later.

The phrase "world's most expensive aquarium" appears in *A Few Good Men from Univac*, p. 167. Both the "love seat" and the "aquarium" are mentioned in a *Pioneer Press* article (December 7, 1981).

Extensive interviews with Seymour Cray appeared in the *St. Paul Pioneer Press* dated (December 7, 1981) and the *Minneapolis Star* (December 22, 1981).

"For three years.... price is too high," quote comes from the *Pioneer Press* (May 20, 1982).

Background information on Steve Chen can be found in *The*

Supermen. Analysis of the product he developed was reported in *Pioneer Press* articles (April 22, April 27, and May 20, 1982). The April 27 article mentions an analyst who describes Chen's work as "the first... to emerge without the aid of engineer of Seymour Cray." His product leadership led to many new commercial customers as mentioned in the 1982 and 1983 annual reports. Rollwagen's remark, "[Seymour] was very gracious," was reported in *BusinessWeek* (April 30, 1990).

The "blockbuster announcement" quote is mentioned in a *Pioneer Press* article (Sept. 14, 1982) that describes the price cuts and the competition. Price reductions creating a "new price performance curve," with products ranging from $4 to $12 million, are also mentioned in the 1982 annual report.

The "80% of the market" quote comes from a Kidder Peabody analyst reported in the *Pioneer Press* (December 16, 1985).

Steve Chen's cancelled development project is described in *BusinessWeek* (April 30, 1990) and in Cray Research's 1987 annual report.

I obtained personal information about Les Davis in an interview with him and from employees who worked with him.

Villafana saying "I understand the project" was reported in the *Star Tribune* (Sept. 5, 1987). There was a follow-up report in the *Pioneer Press Dispatch* (Sept. 23, 1987).

Chen's partnership with IBM is mentioned in various sources including the *Pioneer Press* (May 18, 1988) and the *Star Tribune* (May 16, 1989).

Rollwagen's comments come from a personal interview. The "It isn't working" conversation appears in *The Supermen,* page 161. The "Stunning development" quote from Gary Smaby, plus Control Data losses of $238 million at ETA Systems, and comments on U.S. government requirements, and Japanese competition, can be found in *Time* magazine (May 29, 1989). "Embarrassment of technological riches" comment and "country dominance" comment appear in a *Pioneer Press Dispatch* article (May 16, 1989). Details on the company break-up can be found in a company news release, dated May 15, 1989.

A Control Data Historical Timeline found online says the write-off associated with ETA Systems was $490 million; the book *The Supermen* says $400 million. The company's annual report for 1989 says $335 million, and this is considered the most authoratative source."Seymour Cray's comments about "revolution in the industry" are found in the *Pioneer Press Dispatch* (May 20, 1987).

Potential supercomputer competition from Japanese companies Nippon Electric Co., Hitachi, and Fujitsu are mentioned in the *Star*

Tribune (August 25, 1986). Selling a competitive machine at "half" the purchase price is discussed in the *Star Tribune* (January 4, 1986).

A Control Data agreement to use Cray computers is reported in *Star Tribune* (May 17, 1989).

Seymour's second marriage and personal transformation are described in "The Genius" and a *Pioneer Press* article (December 7, 1981).

Competition from "parallel-processing" is described in *Business-Week* (April 30, 1990). Information on Seymour Cray's final days taken from *The Supermen*, p. 219.

Chapter 11: Conclusion

Cray Research's revenue decline and loss for 1995 was reported in the *Star Tribune* (October 25, 1995). The sale of Cray Research to Silicon Graphics, Inc., is described in the company's 10K filing with the Securities and Exchange Commission for 1995, available at the Hill Reference Library in St. Paul. *Corporate Report Fact Book, 1997*, p. 660, describes the sale as a combination of cash and stock valued at $736 million. Current revenues and stock history for Cray, Inc. can be found in the *Value Line Investment Survey, Small and Mid-Cap Edition.*

Control Data peripheral products activity at $1 billion is mentioned in *A Few Good Men from Univac*. Specific Control Data ventures are mentioned in *The Eye for Innovation* by Robert M. Price.

Former employees recall Control Data's "tomato hydroponics" project. Concerning "cars for convicts," the author talked to four former employees and received various accounts. Dan Moran remembered the liquor store holdup, and so did Gene Baker. Norbert Berg remembered that $40,000 was stolen but the others did not. James Stathopoulos also provided observations. All were in a position to have some knowledge of the activities.

Companies saved from unwanted takeover were American Hoist & Derrick in 1982 and Comserv Corp. a few years later, as reported by the *Minneapolis Star Tribune* (January 13, 1986).

Revenues, earnings, losses, dividends, employment, and related numbers all came from Control Data annual reports, available in the Hill Reference Library, St. Paul. Those reports also provide quotes from company officers and information on various promotions and retirements. Stock prices come from *Corporate Report Fact Books.*

"Social Initiatives" were only 7 percent of corporate revenues in a quote attributed to William Norris by Albert Eisele in a *MinnPost.com* article dated June 16, 2009.

The term "foo foo" is mentioned in *Minneapolis Star Tribune* articles by Dick Youngblood (January 13, 1986, and October 2, 1989). The later article quotes an article published in *Financial World* magazine, by Jagannath Dubashi, entitled "The Do-Gooder." The reference to "eccentric corporate do gooder" in *Inc.* magazine was mentioned in an article in *Minnpost,* June 16, 2009, by Albert Eisele, a former Control Data employee. Eisele, by the way, took exception to the remark.

Quotes from Ed Orenstein and Larry Perlman at the company's fiftieth anniversary celebration were reported in an article about the event by Steve Alexander that appeared in the *Minneapolis Star Tribune* (October 12, 2007).

Details on the initial acquisition of Control Data Systems for $20.25 a share come from the 1998 Corporate Report Fact Book, p. 551. The subsequent acquisition by BT Global Services, and renaming the company Syntegra, was reported in *BusinessWire* (December 1, 1999).

Ceridian was acquired by Thomas H. Lee Partners and Fidelity National Financial in November 2007 and continues to function as a private company.

Final computation of the Control Data parts was done by Jack Karnowski, former financial executive, given in an interview.

Information on private investor Ray Kraushur comes from *Moneymaking.*

The notion that "76 firms" evolved from Univac is found in the book *A Few Good Men from Univac.* The evolution of companies from Control Data is also mentioned in *The Eye for Innovation.*

The Norm Dann quote is taken from *Pioneers of the Medical Device Industry in Minnesota, An Oral History Project,* p. 251.

Medtronic, the world's "largest device company," quote can be found in the *Star Tribune* (October 21, 2010). Financial information on various medical companies is taken from *Value Line Investment Survey* (large companies). Ownership changes over the years for Cardiac Pacemakers are taken from a *Star Tribune* article (March 29, 2012).

The reference to 350 medical technology companies appears in the *Star Tribune* Business Section (June 19, 2012).

The conversation at the 2011 Medtronic annual meeting actually took place in the author's presence. The numbers were verified from company documents.

Index

Donald M. Hall grew up in St. Cloud, Minnesota, attended St. John's University in Collegeville, Minnesota, and Marquette University in Milwaukee, graduating in 1959. He worked at Control Data before moving on to a career in the brokerage industry. Now retired, he lives with his wife, Marion, in Minneapolis.

Made in the USA
Middletown, DE
23 April 2015